The JCT 2011 Building Sub-contracts

The JCT 2011 Building Sub-contracts

Peter Barnes

MSc, FCIOB, FCIArb, MRICS, MICE, MCInstCES

and

Matthew Davies

BSC (Hons), LLB (Hons), LPC, DipAdj, MRICS

WILEY Blackwell

This edition first published 2016
© 2016 JohnWiley & Sons Ltd

Registered office
JohnWiley & Sons, Ltd, The Atrium, Southern Gate, Chichester,West Sussex, PO19 8SQ,
United Kingdom.

Editorial offices
9600 Garsington Road, Oxford, OX4 2DQ, United Kingdom.
The Atrium, Southern Gate, Chichester,West Sussex, PO19 8SQ, United Kingdom.

For details of our global editorial offices, for customer services and for information about
how to apply for permission to reuse the copyright material in this book please see our
website at www.wiley.com/wiley-blackwell.

The right of the author to be identified as the author of this work has been asserted in
accordance with the UK Copyright, Designs and Patents Act 1988.

Library of Congress Cataloging-in-Publication Data applied for.

A catalogue record for this book is available from the British Library.

ISBN: 9781118655634

Wiley also publishes its books in a variety of electronic formats. Some content that appears
in print may not be available in electronic books.

Cover image: peisen zhao/Getty Images

1   2016

To our wives, Carol and Karen, and our families, with our grateful thanks for all the help and support given to us whilst writing this book, without which help and support we would not have been able to do so.

Contents

Preface

The construction industry is almost entirely dependent upon sub-contractors, and on nearly all construction projects a vast majority of the work is carried out by sub-contractors.

Despite this, traditionally, the contract terms relating to sub-contractors have not been given the same consideration as the contract terms in the employer/main contractor relationship. Possibly because of this, there is a clear shortage of books that deal with contract law in the context of sub-contracts.

However, the modern construction industry is fully aware of the importance of sub-contractors and understands that sub-contract terms must be given as much consideration as any other terms. Often the obligations and liabilities of sub-contractors are a vital part of the overall contractual chain leading from the employer/purchaser passing through the contractor and passing down to the sub-contractors and the sub-subcontractors.

It is clear that the JCT recognises the importance of sub-contract terms, and in its 2011 suite of contracts and sub-contracts, it has published a large range of sub-contract forms to be used with many of the current standard forms.

The authors of this book have already written about the sub-contracts in use under the JCT 2005 contracts, and this book now deals with all of the major sub-contracts in relation to the JCT 2011 contract forms.

It is expected that this book will principally be used by sub-contractors and main contractors. However, it will also be of particular interest to other construction professionals and lawyers who need to have an understanding of the contractual relationship and the allocation of risk between main contractors and sub-contractors under the JCT sub-contract forms.

Both of the authors of this book have had wide and direct experience in contractor/sub-contractor relationships and have also had extensive experience in avoiding and resolving disputes in those relationships. Consequently, between us, we have encountered nearly every type of problem that can occur in a contractor/sub-contractor relationship, and we have attempted to interweave some of that knowledge and experience into the text of this book.

This book obviously cannot take into consideration amendments to the standard sub-contract forms that may be made by main contractors. However, it is considered that this book will provide a very useful guide as to the allocation of risk between the parties that exists in the unamended form and also will help both contractors and sub-contractors to understand the possible effects that amendments made to the text of the standard sub-contracts will have on the parties' respective rights and obligations.

Peter Barnes and Matthew Davies
August 2015

1 Background and Introduction

1.1 The Joint Contracts Tribunal (JCT)

The Joint Contracts Tribunal (JCT) was established in 1931 and for over 80 years has produced standard forms of contracts, guidance notes and other standard documentation for use in the construction industry. In 1998, the JCT became incorporated as a company limited by guarantee and commenced operation as such in May 1998.

Currently, JCT forms require the agreement of seven constituent bodies before they are issued by the JCT. Those bodies are:

- The British Property Federation
- The Contractors Legal Group
- The Local Government Association
- The National Specialist Contractors Council
- The Royal Institute of British Architects
- The Royal Institution of Chartered Surveyors
- The Scottish Building Contract Committee

The above listed bodies are intended to be reasonably representative of the interests across the construction industry, namely, the employers, the consultants, the contractors and the sub-contractors, and the JCT sub-contract forms are naturally a reflection of these competing interests.

1.2 Sub-contracting

The regular position is that an employer contracts with a contractor, and the contractor contracts separately and independently with each of his sub-contractors.

The key point in respect of the above relationships is that, although the term 'sub-contract' is used in respect of the contract between the contractor and the sub-contractor, in *all* of the above cases, a *contract* is formed between two parties only (i.e. a contract is formed between an employer and a main contractor; a contract is formed between a sub-contractor and a main contractor).

With that in mind, it would be useful, therefore, to understand some basic principles of contract law.

The JCT 2011 Building Sub-contracts, First Edition. Peter Barnes and Matthew Davies.
© 2016 John Wiley & Sons, Ltd. Published 2016 by John Wiley & Sons, Ltd.

Most aspects of the law of contract are set down in case law; however, there are some notable exceptions where provision is made in statute (e.g. the Sale of Goods Act 1979, the Unfair Contract Terms Act 1977 and the Supply of Goods and Services Act 1982).

Because of the nature of this book, the basic principles of contract law, as provided at section 1.3 can naturally be dealt with in outline only.

1.3 The formation of contracts and sub-contracts

There are many definitions of a contract, but in simple terms, it can be considered as being: 'an agreement which gives rise to obligations which are enforced or recognised by law'. Under English law, only the actual parties to a contract can acquire rights and liabilities under the contract. This is known as 'privity of contract'.

In respect of a main contract situation, the practical consequences of the doctrine of privity of contract are twofold:

- the main contractor carries responsibility for a sub-contractor's work, etc., so far as the employer is concerned; and
- the employer cannot take direct action in contract against the sub-contractor, unless there is a separate contract between the employer and the sub-contractor.

The effect that the Contracts (Rights of Third Parties) Act 1999 has upon this position in respect of the JCT sub-contracts considered in this book is dealt with later within this book.

The essence of any contract is *agreement*. In deciding whether there has been an agreement, and what its terms are, the court looks for an *offer* to do or to forbear from doing something by one party and an unconditional *acceptance* of that offer by the other party, turning the offer into a promise.

In addition, the law requires that a party suing on a promise must show that he or she has given *consideration* for the promise, unless the promise was given by *deed*.

Further, it must be the intention of both parties to be legally bound by the agreement, and the parties must have the capacity to make a contract, and any formalities required by law must be complied with. Finally, there must be sufficient *certainty of terms*.

1.3.1 Offer

An offer is a statement by one party of a willingness to contract on definite stated terms and intended to be binding, provided that these terms are, in turn, unequivocally[1] accepted by the party or parties to whom the offer is addressed.

---

[1] The acceptance must be unqualified; it must 'mirror' the offer i.e. it must accept all of the terms being offered without changing or seeking to add to, vary or amend the terms of the offer being made.

There is generally no requirement that the offer be made in any particular form; it may be made orally, in writing or by conduct. Of course, if a dispute arose in the future, then it would be beneficial for the offer to be in writing.

In whichever form an offer is made, it must be sufficiently definite to be capable of resulting in a contract if accepted. Its terms and conditions must be clear and unequivocal, and it must be made with the intention that it is to become binding as soon as it is accepted by the person to whom it is addressed. In this context, a person includes a corporation because, in law, a corporation is a legal person; that is to say, a corporation is regarded by the law as a legal entity quite distinct from the person or persons who may, for the time being, be the member or members of the corporation.

Putting the above into context, it is generally the case that when the sub-contractor submits his estimate (i.e. his tender), this is an offer which the contractor can either accept or reject.

With the above in mind, it must be noted by sub-contractors that the submission of a tender does not (normally) conclude a contract. Therefore, the preparation of a tender in response to a tender enquiry (which would, in the normal course of events, become an offer when submitted) may involve the sub-contractor in (sometimes considerable) expense, but the cost of tender preparation is not normally recoverable as a discrete cost. Obviously, the cost of tender preparations is included within the head office overhead percentage that is added by sub-contractors onto their tenders, and the tender preparation costs so incorporated are therefore recovered by sub-contractors when their tenders are successful.

1.3.2   Acceptance

For agreement to be reached, there must be a clear and unequivocal acceptance of a clear and unequivocal offer. The acceptance must be unqualified;that is, as noted earlier, it must 'mirror' the offer.

Therefore, if in a tender enquiry, a sub-contractor was required to use Welsh slates but submitted his tender on the basis of using Spanish slates, and the contractor, upon receiving the sub-contractor's tender, accepted the sub-contractor's tender without qualification, then the contract, when formed, would be on the basis of the terms and conditions which formed part of the tender (i.e. based on using Spanish slates rather than Welsh slates).

In such a situation, if a future dispute arose, the contractor would not be able to rely on the terms and conditions forming part of the tender enquiry (i.e. that Welsh slates were required) because those terms and conditions would not form part of the contract between the parties.

As a general rule, silence does not constitute acceptance[2]; neither does inactivity.

Given this, the general rule is that an acceptance has no effect until it is communicated (either in writing or orally) to the party making the offer. The main

---

[2]*Felthouse* v. *Bindley* (1862) aff'd (1863).

reason for this being that it could cause hardship to the party making the offer if he or she were bound without knowing that his or her offer had been accepted.

Another rule is that refusal of an offer puts an end to that offer.[3] Hence, if a contractor rejects a sub-contractor's offer (i.e. his submitted tender), it is, in legal parlance, extinguished and is no longer legally capable of being accepted.

An offer can be withdrawn at any time before it is accepted.[4] This rule even applies where an offer is stated to be open for a fixed time. Therefore, if a sub-contractor submitted a tender (or a quotation) and stated that it remained open for acceptance within 30 days, there would be nothing to prevent the sub-contractor from withdrawing that offer in a period less than 30 days (unless the express terms of the sub-contract stated otherwise).

If an offer is stated to be open for a fixed time, then it cannot (without the agreement of the party making the offer) be accepted after that time.

However, if no time is stated in the offer, then the offer is taken to lapse after a reasonable time. The word 'reasonable' is, of course, open to interpretation, but is based on the facts of the particular case.

In certain circumstances, depending on the facts of each particular case, acceptance may be made by conduct[5] (e.g. allowing possession, making payments in line with the agreed terms) or by performance[6] (e.g. by carrying out and completing the sub-contract works).

1.3.3   Counter-offer

If the acceptance does not clearly and unequivocally accept the offer (e.g. it seeks to add to or vary the terms contained in the offer), then it is, simply, a counter-offer (not an acceptance), and this simply destroys the original offer.

A counter-offer has the same status as an offer in the formation of a contract, and consequently, a counter-offer must be clearly and unequivocally accepted before agreement has been reached.

Counter-offers should be distinguished from requests for information, which will not necessarily amount to a counter-offer.[7] Care must be taken when requesting further information to ensure this request is not construed, in fact, as being a counter-offer.

In respect of construction works, in particular, there are frequently negotiations during which a whole series of counter-offers are made between the respective parties to each other before an acceptance is finally made by one party to the other party. This may be because the parties are negotiating about the terms or because they are each trying to impose their own terms.

This latter situation is often called the 'battle of the forms', and variants of this battle of the forms are at the base of many disputes between contractors and sub-contractors.

---

[3]*Hyde* v. *Wrench* (1840) 3 Beav 334.
[4]*Dickinson* v. *Dodds* (1876) 2 Ch D 463.
[5]*G Percy Trentham* v. *Archital Luxfer Ltd* (1992) 63 BLR 44, CA.
[6]*Brogden* v. *Metropolitan Railway Co* (1877) 2 App Cas 666.
[7]*Stevenson, Jaques & Co* v. *McLean* (1880).

1.3.4 The 'battle of the forms'

The expression 'battle of the forms' refers to an offer followed by a series of counter-offers where each party successively seeks to stipulate different terms, often based on their own standard printed terms. It is sometimes extremely difficult to determine whether or when a concluded contract came into existence, particularly where no formal contract is ever signed, and it is then left to the courts to determine the matter.

Clearly, this is far from satisfactory, and it makes sense to ensure that the sub-contract terms are agreed and recorded in writing to avoid the need to rely on a court interpreting the intention of the parties from an analysis of the various exchanges of communication.

1.3.5 Agreement

As noted at the beginning of section 1.3 the essence of any contract is agreement.

The test for the existence of an agreement is objective rather than subjective. In other words, the existence is tested on the facts, rather than on what may have been perceived to be the intention of the parties.

The principal justification for the adoption of this test is the need to promote certainty.

An agreement is reached either:

- when a statement of agreement is signed; or
- when one party makes an unambiguous offer capable of being accepted, and the other party accepts it unequivocally.

1.3.6 Certainty of terms

Even if there is clearly agreement through offer and acceptance, a contract may fail to come into existence because of uncertainty as to what has been agreed[8].

Although it has been found that 'the parties are to be regarded as masters of their contractual fate in determining what terms are essential'[9] and 'it is for the parties to decide whether they wish to be bound and, if so, by what terms, whether important or unimportant',[10] it is generally considered that agreement as to parties, price, time and description of works is normally the minimum necessary to make the contract commercially workable.

Silence by the parties as to either price or time may not alone prevent a contract coming into existence, for if the other essential terms are agreed, a reasonable price or time for completion may be implied by the Supply of Goods and Services Act 1982.

---

[8]*Scammell v Ouston* (1941) 1 ALL ER 14
[9]*Pagnan v. Feed Products* [1987] 2 Lloyds Rep 601, CA.
[10]*Mitsui Babcock Energy Ltd v. John Brown Engineering Ltd* (1996) CILL 1189.

1.3.7   Consideration

Other than where a contract is executed as a deed (see further commentary later in this book at chapter 2, section 2.6.1), an agreement requires consideration to be exchanged between the contracting parties before it becomes binding.

The classic definition of consideration was expressed in *Currie* v. *Misa* (1875)[11] in the following terms:

> 'a valuable consideration, in the sense of the law, may consist either in some right, interest, profit or benefit accruing to the one party, or some forbearance, detriment, loss or responsibility given, suffered or undertaken by the other'.

In the ordinary building sub-contract situation, the consideration given by the contractor is the price paid or the promise to pay, and the consideration given by the sub-contractor is the carrying out of the works or the promise to carry them out.

The rules which make up the doctrine of consideration may be divided into three categories:

1. Consideration must be sufficient but it need not be adequate.
2. Past consideration is not good consideration (i.e. the general rule is that the consideration must relate to future works. Any acts already performed that are the subject matter of the agreement do not, generally, constitute good consideration and are unenforceable).
3. Consideration must move from the party to whom the promise has been made.

As noted above, consideration is not required in the case where a contract is executed as a deed. Contracts, other than made by deed, are termed simple contracts, whether made orally or in writing.

1.4   Standard forms of contract and sub-contract

There are obvious benefits in using standard forms of contract and sub-contract. These include:

- There is no need to produce (and incur the legal costs of producing) *ad hoc* contracts and sub-contracts for every project.
- There is a degree of certainty regarding the interpretation of the clauses of the contract (particularly those standard forms that have been in existence for some time and where some of the more important clauses may have been tested in the courts).
- The parties know (with reasonable certainty) the consequences of various possible courses of action.

---

[11]*Currie* v. *Misa* (1875) LR 10 Ex 153.

Standard forms of contract (perhaps less so of standard forms of sub-contract) can be traced back to the nineteenth century (if not earlier). However, it appears to be a fairly recent phenomenon, largely emanating from the Latham Report (*Constructing the Team*, 1994), that standard forms and sub-contracts are now seen as having a dual purpose. These are:

- To set out the rights and obligations of the parties.
- To ensure that the risk of the project is allocated to the party that can best manage that risk.

In this regard, there have been great steps taken, particularly in respect of main contracts, but it is questionable whether that same progress has been made in respect of sub-contracts.

It is still prevalent that main contractors produce their own sub-contract forms, or, more commonly, issue their own amendments to the standard sub-contract forms.

Perhaps unfortunately, under the JCT main contract forms, there is nothing to prevent this from happening in the future, and because of this, it is considered by some that main contractors under JCT contracts should be obliged to sub-let works using JCT sub-contracts.

In 2011, the JCT authorised the publication of an entirely new suite of contracts and sub-contracts, etc.

Amongst the sub-contracts that have been issued are the following sub-contracts:

- Standard Building Sub-contract
- Standard Building Sub-contract with sub-contractor's design
- Design and Build Sub-contract
- Major Project Sub-contract
- Intermediate Sub-contract
- Intermediate Sub-contract with sub-contractor's design
- Intermediate Named Sub-contract
- Minor Works Sub-contract with sub-contractor's design
- Short Form of Sub-contract
- Sub-subcontract
- Management Works Contract (a sub-contract under the Management Building Contract)

1.5   The JCT Sub-contracts dealt with within the chapters of this book

The following sub-contracts are dealt with within the chapters of this book:

- Standard Building Sub-contract
- Standard Building Sub-contract with sub-contractor's design
- Design and Build Sub-contract

- Intermediate Sub-contract
- Intermediate Sub-contract with sub-contractor's design

1.5.1 The standard building sub-contract

The Standard Building Sub-contract can be used where the main contract is the 2011 issue of any of the three versions of the Standard Building Contract 2011, that is, With Quantities (SBC/Q), With Approximate Quantities (SBC/AQ) or Without Quantities (SBC/XQ).

The Standard Building Sub-contract is for use where the Sub-contractor is not required to design any of the Sub-contract Works.

The Standard Building Sub-contract can be used when the Main Contract Works are to be carried out in Sections, and can be used either where the Sub-contractor is to be paid a Lump Sum, adjustable for variations, etc., or where there is an agreed Tender Sum but the sub-contract works are to be subject to complete re-measurement.

The Standard Building Sub-contract, referred to as the SBCSub, comprises two documents:

- the Agreement (SBCSub/A), and
- the Conditions (SBCSub/C).

1.5.2 The standard building sub-contract with sub-contractor's design

The Standard Building Sub-contract with sub-contractor's design can be used where the main contract is the 2011 issue of any of the three versions of the Standard Building Contract 2011, that is, With Quantities (SBC/Q), With Approximate Quantities (SBC/AQ) or Without Quantities (SBC/XQ).

The Standard Building Sub-contract with sub-contractor's design is for use where the Sub-contractor is required to design all or part of the sub-contract works.

The Standard Building Sub-contract can be used when the Main Contract Works are to be carried out in Sections and can be used either where the Sub-contractor is to be paid a Lump Sum, adjustable for variations, etc., or where there is an agreed Tender Sum but the sub-contract works are to be subject to complete re-measurement.

The Standard Building Sub-contract, referred to as the SBCSub/D, comprises two documents:

- the Agreement (SBCSub/D/A), and
- the Conditions (SBCSub/D/C).

1.5.3 The design and build sub-contract

The Design and Build Sub-contract is only suitable for sub-contracts where the main contract is the 2005 edition of the JCT Design and Build Contract.

Within this book, the JCT Design and Build Sub-contract will be referred to as DBSub (the designation given to it by the JCT).

This sub-contract can be used even where the sub-contractor is not to carry out any design work at all, and can be used for either lump sum or re-measurement contracts.

The sub-contract can be used when the main contract works are to be carried out in Sections as detailed in the main contract.

The sub-contract comprises two documents:

- the Agreement (denoted by the JCT and in this book as DBSub/A), and
- the conditions (denoted by the JCT and in this book as DBSub/C).

1.5.4 The intermediate sub-contract

This sub-contract is suitable for sub-contracts where the main contract is the JCT Intermediate Building Contract *or* where the main contract is the JCT Intermediate Building Contract with contractor's design; but even where the main contract form with contractor's design is used, the Intermediate Sub-contract is only for use when the sub-contractor is not liable for design.

Within this book, the Intermediate Sub-contract will be referred to as ICSub (the designations given to it by the JCT).

This sub-contract can be used when the main contract works are to be carried out in Sections as detailed in the main contract.

This sub-contract cannot be used where the sub-contractor is to be named or where the sub-contractor is to carry out any design work, but can be used for either lump sum or re-measurement contracts.

The sub-contract comprises two documents:

- the Agreement (denoted by the JCT and in this book as ICSub/A), and
- the Conditions (denoted by the JCT and in this book as ICSub/C).

1.5.5 The intermediate sub-contract with sub-contractor's design

This sub-contract is suitable for sub-contracts where the main contract is the JCT Intermediate Building Contract with contractor's design.

Within this book, the Intermediate Sub-contract with sub-contractor's design will be referred to as ICSub/D (the designations given to it by the JCT).

This sub-contract can be used when the main contract works are to be carried out in Sections as detailed in the main contract.

This sub-contract is to be used where the sub-contractor is to design all or part of the sub-contract works, and can be used for either lump sum or re-measurement contracts.

This sub-contract is not to be used where the sub-contractor is to be named. The sub-contract comprises two documents:

- the Agreement (denoted by the JCT and in this book as ICSub/D/A), and
- the Conditions (denoted by the JCT and in this book as ICSub/D/C).

1.6 The JCT Sub-contracts not dealt with in the chapters of this book

The following sub-contracts are not dealt with within the chapters of this book but in outline as follows:

- Major Project Sub-contract
- Intermediate Named Sub-contract
- Minor Works Sub-contract with sub-contractor's design
- Short Form of Sub-contract
- Sub-subcontract
- Management Works Contract (a sub-contract under the Management Building Contract)

1.6.1 The major project sub-contract

The Major Project Sub-contract (the 'MPSub') is only suitable for sub-contracts where the main contract is the JCT Major Project Construction Contract.

This sub-contract does not have a separate Agreement document, but relies on a section within the sub-contract called the Sub-contract Particulars to provide the information that would normally be included in the appendix to a pre-2005 JCT form.

The MPSub can be used whether or not the sub-contractor is responsible for design, and can be used for either lump sum or re-measurement contracts.

The sub-contract works on site commence on notice from the Contractor, and although the sub-contract contains no specific provisions for the completion of the sub-contract works in sections, a similar effect can be achieved by setting out within the sub-contract particulars specific binding dates for particular items.

Interim payments are to be made monthly, but through the use of the pricing document, it is possible to adopt a range of options for the payment of the sub-contract sum, including interim valuations, stage payments and/or scheduled payments. The sub-contract does not provide for retention to be held from any payment.

The contractor may require changes to either the requirements or the proposals (if applicable), or in the manner in which the sub-contractor undertakes the sub-contract works. The financial consequences of changes are intended to be determined on an all-inclusive basis (i.e. they are to include any loss and expense that may be incurred as a result of the change) and are intended to be agreed in advance.

There is no provision for arbitration; disputes may be resolved by mediation, adjudication or litigation.

1.6.2 The intermediate named sub-contract

This sub-contract is suitable for sub-contracts where the main contract is the JCT Intermediate Building Contract *or* where the main contract is the JCT Intermediate Building Contract with contractor's design.

As its name implies, this sub-contract is to be used where a sub-contractor is named (by the employer) to carry out the sub-contract works.

The Intermediate Building Contract and the Intermediate Building Contract with contractor's design are the only contracts in the JCT suite of contracts that allow for the naming of a sub-contractor. It should be noted that a named sub-contractor does not have the status that a nominated sub-contractor had under the JCT 98 Main Contract Form, and after a named sub-contractor is appointed, he or she effectively simply becomes a domestic sub-contractor.

This sub-contract can be used when the main contract works are to be carried out in Sections as detailed in the main contract.

It is not suitable for any sub-contract work that forms a part of the contractor's designed portion, but can be used where the sub-contractor is required to design all or part of the sub-contract work. In other words, any design carried out by a named sub-contractor cannot be design that forms part of the contractor's designed portion.

This sub-contract can be used for either lump sum or re-measurement contracts.

The Sub-contract comprises four documents:

■ the invitation to tender (ICSub/NAM/IT),
■ the tender (ICSub/NAM/T),
■ the agreement (ICSub/NAM/A), and
■ the conditions (ICSub/NAM/C).

The process of a named sub-contractor being appointed starts with an invitation to tender being issued to the proposed sub-contractor normally by the employer or his or her agent (ICSub/NAM/IT).

Upon receipt of the invitation to tender form, the proposed named sub-contractor provides his or her tender to the employer's/agent using the form ICSub/NAM/T. The sub-contractor's tender is to provide, amongst other things, the sub-contractor's price and the basis upon which that price is supplied, the required programme details, and any additional attendance items required.

The main contractor first becomes aware of a prospective named sub-contractor when the sub-contractor's tender (and the invitation to tender) is included in:

■ the original tender enquiry to the main contractor,
■ an architect's/contract administrator's instruction for the expenditure of a provisional sum under the main contract, or
■ in an architect's/contract administrator's instruction for the expenditure of a provisional sum naming a replacement named sub-contractor.

The Intermediate Building Contract and the Intermediate Building Contract with contractor's design set out the various procedures that must be followed by a contractor in respect of the appointment of a named sub-contractor (and those procedures deal with the situation where a contractor objects to the proposed named sub-contractor or the proposed terms of that sub-contractor).

It is for the main contractor and the prospective named sub-contractor to agree upon the actual terms of the sub-contract. Those terms will, of course, be based upon the ICSub/NAM/IT and the ICSub/NAM/T, and the main contractor therefore, in effect, accepts the named sub-contractor's tender, but there may be amendments and/or additions agreed to the terms contained in these documents. Any such amendments and/or additions are to be noted in the numbered documents referred to under the second recital of the ICSub/NAM/A, and listed under the numbered documents section of the contract particulars of the ICSub/NAM/A.

When a sub-contract is entered into between a main contractor and a named sub-contractor, the sub-contract form used is the ICSub/NAM. When this occurs, the named sub-contractor becomes, for all intents and purposes, a normal domestic sub-contractor of the main contractor, and the main contractor has the same liability in respect of the named sub-contract works (sub-contract works is a defined term in the sub-contract) as it would for the works of a normal domestic sub-contractor. The only exception to this position is that if the contractor becomes aware of events that may lead to the termination of a named sub-contractor's employment, the contractor must notify the architect/contract administrator accordingly. If termination subsequently takes place because of the default or insolvency of a named sub-contractor, then, with the architect/contract administrator's consent, the contractor is entitled to some relief from the financial consequences of that event.

The use of the named sub-contractor procedure is intended primarily for work involving a design input by the named sub-contractor, and this may be the case even where the installation of the specialist work is not of a complex nature. Under paragraph 11.1 of schedule 2 of the main contract, the contractor is in such cases expressly relieved of responsibility to the employer for defects in the named sub-contractor's design of the sub-contract works.

Because there is no contractual link between the employer and the named sub-contractor, and as the contractor is relieved of responsibility for the named sub-contractor's design, the JCT suggests that an employer may consider using an intermediate named sub-contractor–employer agreement (ICSub/NAM/E), which provides a direct contractual link between the employer and the named sub-contractor for use where the named sub-contractor is to carry out design work or is to procure or fabricate materials or goods prior to the letting of the main contract.

1.6.3 The minor works sub-contract with sub-contractor's design

This sub-contract is suitable for sub-contracts where the main contract is the JCT Minor Works Building Contract with contractor's design.

This sub-contract is for use with small sub-contract packages with a straightforward content and a low risk involved, and is only to be used where the sub-contractor is required to design all or part of the sub-contract works.

This sub-contract does not have a separate Agreement document, but relies on the Recitals and the Articles to provide the information that would normally be included in an appendix to a pre-2005 JCT form.

Provision for brief entries identifying the main contract conditions are included in the first recital.

Articles 2 and 3 deal with the date for commencement and the period for completion, respectively. The guidance notes state that any information beyond this, for example, site working times or any working restrictions, should, if necessary, be set out in the pricing document or further documents inserted in the second recital.

Details of the main contract particulars and the amendments to those particulars should be provided under the third recital.

Despite the sub-contract's brevity, it is in the sub-contractor's interest to ensure that the sub-contract is provided with clear, accurate and sufficient information in respect of the main contract and any matters relevant or affecting the sub-contract works, especially given the obligations placed on the sub-contractor to comply with the main contract provisions at clause 7.

The first interim payment becomes due not later than 1 month after commencement of the sub-contract works on site. Subsequent interim payments are due at monthly intervals after the first interim payment. The final date for payment is 21 days after the due date.

This sub-contract makes no provision for a sub-contractor to make a payment application in advance of the payment due date, but the sub-contractor is permitted to make a payment application as a default payment notice.

The amount due for payment in respect of interim payments is 95% (or such other percentage as may be set out in the sub-contract documents) of the value of the work properly carried out, together with any applicable loss and/or expense, less the total amount due in any previous payments.

After practical completion of the sub-contract works, the amount due for payment increases to 97.5% (or such other percentage as may be set out in the sub-contract documents).

The value of the works properly executed is determined by reference to the rates and prices in the pricing document or by reference to the sub-contract sum if there are no such rates and prices.

This sub-contract does not envisage that payment will be made for off-site materials.

A payment notice is to be given not later than 5 days after the date on which a payment becomes due. That notice is to specify the amount of the payment to be made and the basis on which that amount was calculated.

A Pay Less notice (if applicable) is to be given not later than 5 days before the final date for payment of an interim payment.

This sub-contract does not require the sub-contractor to provide information to the contractor to enable the contractor to calculate the final sub-contract sum.

Article 1 of the sub-contract conditions makes it clear that the sub-contract sum excludes VAT. Any VAT properly chargeable is to be paid by the contractor to the sub-contractor.

This sub-contract does not contain an express term preserving the sub-contractor's common law rights to claim damages for breach of contract;

however, it is considered that their silence on this point does not affect the sub-contractor's right to bring a claim based on his or her common law rights.

Clause 10.3 restricts the sub-contractor's right to payment of any direct loss and/or expense to circumstances due to the regular progress of the sub-contract works being affected by the compliance with any written variation. Such recovery is also subject to the sub-contractor notifying the contractor of the same as soon as it is 'reasonably practicable'. Under this provision, the contractor is obliged to determine a fair and reasonable amount of direct loss and/or expense.

The limited right to payment of loss and/or expense under the sub-contract may cause some problems in practice, for example, not all instructions may constitute variations. However, if additional costs due to matters other than variations are incurred by the sub-contractor, then these may be pursued as a claim for common law damages in appropriate circumstances.

All variation instructions are to be issued in writing in line with clause 9.1.

If the contractor gives oral instructions, those instructions are to be confirmed by the contractor within two working days of being given (as clause 9.2).

Clause 10.1 requires the sub-contractor to carry out any reasonable variation of the sub-contract works that is instructed in writing by the contractor. Of course, there could be many disputes regarding what a 'reasonable variation' is, and the wording of clause 10.1 is therefore not entirely satisfactory.

Clause 9.1 notes that the contractor is not to issue any instructions affecting the design of the sub-contractor's design portion works without first obtaining the consent of the sub-contractor (consent that the sub-contractor shall not unreasonably withhold).

Clause 10.2 notes that variations shall be valued by the contractor on a fair and reasonable basis, with reference, where available and relevant, to rates and prices in the pricing document (which being any document identified in the second recital of the contract that shows rates and prices).

1.6.4   The short form of sub-contract

This sub-contract is suitable for any sub-contracts (other than for named sub-contractors) where the main contract is a JCT contract. However, it is not suitable where provisions which are fully back to back with a particular main contract are required.

This sub-contract can be used when the main contract works and/or the sub-contract works are to be carried out in Sections, and can be used for either lump sum or re-measurement contracts.

This sub-contract does not have a separate Agreement document, but relies on the Recitals and the Articles to provide the information that would normally be included in an appendix to a pre-2005 JCT form.

This sub-contract is intended for use in respect of small sub-contract packages of straightforward content and with low risk involved. It is not suitable where the sub-contractor is to design any part of the sub-contract works.

1.6.5   The sub-subcontract

This sub-contract is an entirely new concept for the JCT in that it is a tertiary contract. It is intended for use when a sub-contractor wishes to place a sub-contract with his or her own sub-contractor(s), that is, the sub-subcontractor(s).

It is suitable for sub-subcontracts where the main contract is a JCT contract, and can be used with any sub-contract. However, it is not suitable where provisions which are fully back to back with a particular sub-contract are required.

This sub-subcontract can be used when the main contract works and/or the sub-contract works are to be carried out in Sections, and can be used for either lump sum or re-measurement contracts.

This sub-subcontract does not have a separate Agreement document, but relies on the Recitals and the Articles to provide the information that would normally be included in an appendix to a pre-2005 JCT form.

1.6.6   The management works contract (a sub-contract under the management building contract)

The Management Works Contract 2011 is suitable for sub-contracts where the main contract is the JCT Management Building Contract 2011.

A management contractor does not carry out the construction works at the project (he is paid a fee for managing the project and providing necessary site facilities). The management contractor directly engages specialist contractors to execute the construction works at the Project under sub-contracts.

As noted above the JCT publish the JCT Management Works Contract 2011 as a suitable sub-contract where the main contract is the JCT Management Building Contract 2011.

The Management Works Contract 2011 comprises two documents:

- The management works contract agreements (MCWC/A), and,
- The management works contract conditions (MCWC/C).

This sub-contract can be used when the sub-contract works or the Project are to be carried out in Sections. This sub-contract can also be used whether or not the sub-contractor is responsible for design. In terms of payment to the sub-contractor this sub-contract can be used for either a lump sum price (with adjustment for variations, provisional sums) or in the alternative it has the option for payment to be on a re-measurement basis.

The management works contract conditions (MCWC/C) has a familiar JCT style in its layout, and is in 9 sections,

- Section 1 – entitled *"definitions and interpretation"*.
- Section 2 – entitled *"carrying out the works"*.
- Section 3 – entitled *"control of the works"*.
- Section 4 – entitled *"Payment"*.
- Section 5 – entitled *"valuation of the works and variations"*.

- Section 6 – entitled "*injury, damage and insurance*".
- Section 7 – entitled "*termination*".
- Section 8 – entitled "*settlement of disputes*".

There are also 7 schedules appended dealing with various matters, such as Design submission procedure (schedule 1), acceleration quotation and variation procedure (schedule 2), third party rights (schedule 3) etc.

As noted above, the sub-contract sets out the payment mechanism (terms, requirements, obligations, rules, etc) at section 4 of the management works contract conditions, dealing with payment matters such as (but not limited to), for example:

- The method of calculating payment whether by:
  - The Adjustment basis (where Article 3 A of the MCWC/A is selected), including the method of calculating the Final Works Contract Sum.
  - The remeasurement basis (where Article 3 B of the MCWC/A is selected), including the method of calculating the Final Works Contract Sum.
- Payments and notices
  - Interim payments
  - Final payment
- Construction Industry Scheme
- Retention
- Loss and expense.

2 The Sub-contract Agreement

2.1 Introduction

People frequently refer to the terms and conditions of a contract or a sub-contract.

Using this terminology, the terms of a sub-contract would normally constitute the particular insertions made by the parties to the sub-contract into the sub-contract, whereas the conditions would already be part of the standard published form.

In respect of some of the sub-contracts under consideration in this book,[1] there is a separate agreement document where the terms are inserted by the parties, whilst for others,[2] there is only one sub-contract document and the terms are inserted into the document itself in the same way as would have been the case for an old-style JCT appendix. In the cases where there are two separate sub-contract documents, there is a clause in the sub-contract (e.g. clause 1.3 under the SBCSub/D/C) which makes it clear that the sub-contract is to be read as a whole.

Whatever is inserted as a term into the executed version of the sub-contract will be an express term agreed between the parties. The law will understand these express terms to be freely agreed between the parties, and any argument put forward at a later date that the express terms were not freely agreed is unlikely to meet with any success.

Also, the terms will be the only express terms applicable to the sub-contract. Therefore, whatever may have been discussed and/or written about before the execution of the sub-contract will not form part of the express terms of the sub-contract unless the matter in question is specifically listed within the sub-contract.

---

[1] i.e. the Standard Building Sub-contract; the Standard Building Sub-contract with sub-contractor's design; the Design and Build Sub-contract; the Intermediate Sub-contract; and the Intermediate Sub-contract with sub-contractor's design

[2] i.e. the Major Project Sub-contract; the Intermediate Named Sub-contract; the Minor Works Sub-contract with sub-contractor's design; the Short Form of Sub-contract; and the Sub-subcontract.

The JCT 2011 Building Sub-contracts, First Edition. Peter Barnes and Matthew Davies.
© 2016 John Wiley & Sons, Ltd. Published 2016 by John Wiley & Sons, Ltd.

Consequently, it is important for both the contractor and the sub-contractor that everything that should be referred to and listed within the sub-contract *is* referred to and listed within the sub-contract. The corollary to the above is, of course, that anything that should not be referred to and listed within the sub-contract particulars should *not* be referred to and listed within the sub-contract particulars.

By way of an example, if it was discussed that the sub-contractor was to clear his or her waste materials and rubbish to skips to be provided by the contractor – but this matter was not listed within the sub-contract; then, if this matter became an issue at a later date, the sub-contractor would not have the contractual right to insist that the contractor provide skips for the removal of the sub-contractor's waste materials and rubbish. Conversely, if it was discussed that the sub-contractor would not be charged for the use of mess rooms – but this matter was not listed within the sub-contract; then, if this matter became an issue at a later date, the contractor would have the contractual right to charge the sub-contractor when the sub-contractor used the mess rooms.

The parties need to be aware that the meaning of a sub-contract will be ascertained by a court from the words actually used in that written sub-contract, and a court normally will not hear evidence from the parties as to what they may have meant rather than what the words of the sub-contract say.

This 'rule' of sub-contract interpretation is sometimes known as the 'parole evidence rule', and although the courts are moving away from the use of formal 'rules' towards a more common-sense approach to sub-contract interpretation,[3] it would still be generally futile (if a dispute arose) for a party to rely on any negotiations leading up to a written sub-contract (if those negotiations are not recorded in some way within the written sub-contract) in an attempt to change in some way the plain wording of the sub-contract. This point can represent a trap for contractors and sub-contractors alike who may wish to refer to the sub-contract negotiations in an attempt to show that the alleged intention of the parties was different to the actual wording of the sub-contract.

It is extraordinary the number of times that contractors and sub-contractors take great care over agreeing terms that they both feel comfortable with, only to fail to properly record those agreements in the sub-contract. Experience shows that when disputes arise, people tend to have very selective memories; therefore, and particularly as a court or a tribunal will generally only be interested in clear evidence, if a term has been agreed, it really must be recorded in the sub-contract if a party wishes to rely upon that term in the future.

Against this background, in this chapter, the following matters will be dealt with, using the sub-contract clause references, etc. contained in the SBCSub-/D/A or SBCSub/D/C (as appropriate). Whilst it is clearly beyond the scope of

---

[3] See *Investors Compensation Scheme Ltd* v. *West Bromwich Building Society* [1997] UKHL 28; *Rainy Sky SA and Others* v. *Kookmin Bank* [2011] UKSC 50.

this book to review every nuance of the other sub-contract forms under consideration, the equivalent provisions (where applicable) within the SBCSub/A or SBCSub/C, the DBSub/A or DBSub/C, the ICSub/A or ICSub/C and the ICSub/D/A or ICSub/D/C are given in Table 2.1. It must be emphasised that before considering a particularly issue, the actual terms of the appropriate edition of the relevant sub-contract should be reviewed by the reader (and/or legal advice should be sought as appropriate) before proceeding with any action/ inaction in respect of the sub-contract in question.

The matters dealt with in this chapter are:

- The Structure of the Sub-contract Agreement
- The Recitals
- The Articles
- The Sub-contract Particulars
- The Attestation Form
- The Schedule of Information
- The Supplementary Particulars

2.2   The structure of the sub-contract agreement

In the context of the SBCSub/D, the Agreement is the SBCSub/D/A, and this incorporates:

- The date that the agreement is made (normally the date that the second party executes the agreement).
- The names and addresses of the parties (i.e. the contractor and the sub-contractor).
  - ○ Normally, the parties' company number should be inserted, and their registered office address should be used. If the contractor or the sub-contractor is a company incorporated outside England and Wales, particulars of its place of incorporation should be inserted immediately before the company number. If the contractor or the sub-contractor is neither a company incorporated under the Companies Act nor a company registered under the laws of another country, the references to a company number and to a registered office address should be deleted.
- Details (by identification only) of the sub-contract works, and the main contract works of which they form a part

This is then followed by:

- Ten Recitals
- Six Articles
- Sub-contract Particulars items 1–17
- Attestation Forms
- A Schedule of Information
- Supplementary Particulars

2.3 The recitals

The recitals set out certain facts as a background to the parties entering into the sub-contract.

2.3.1 First recital

This recital refers to the main contract as described in the schedule of information annexed to the agreement as dealt with later within this chapter.

2.3.2 Second recital

The second recital states that the contractor wishes the sub-contractor to execute the sub-contract works described in the numbered documents.

2.3.3 Third recital

This recital is to be completed by the parties to identify the part of the sub-contract works that the sub-contractor is responsible for in respect of both design and construction. The entry made may be descriptive, or it may refer to the document(s) that more fully describes the relevant part of the sub-contract works. Under the SBCSub/D/A, the entry made in this recital is known as 'the Sub-contractor's Designed Portion'.

2.3.4 Fourth recital

The fourth recital confirms that the contractor has provided documents showing and describing or otherwise stating the requirements of the contractor for the design and construction of the Sub-contractor's Designed Portion. These documents, which may include the Employer's Requirements and the Contractor's Proposals, should be listed under item 17 of the sub-contract particulars as numbered documents; and the said documents are known collectively as the Contractor's Requirements. To avoid future dispute, the documents comprising the Contractor's Requirements should be annexed to the sub-contract agreement and signed or initialled by or on behalf of each party.

2.3.5 Fifth recital

This recital confirms that in response to the Contractor's Requirements (as identified in the fourth recital), the sub-contractor has supplied to the contractor:

- documents showing and describing the proposals of the sub-contractor for the design and construction of the Sub-contractor's Designed Portion. These documents are known collectively as the Sub-contractor's Proposals, and
- cost information in the form of a Sub-contractor's Designed Portion analysis as applicable to the sub-contract sum or the sub-contract tender sum (depending upon which of these options is selected under article 3A or article 3B.)

All documents comprising the Sub-contractor's Proposals and the Sub-contractor's Designed Portion analysis should be listed as numbered documents in the sub-contract particulars (item 17), and to avoid future dispute, these documents should also be annexed to the sub-contract agreement and signed or initialled by or on behalf of each party.

2.3.6 Sixth recital

The sixth recital confirms that the contractor has examined the Sub-contractor's Proposals and, subject to the sub-contract conditions, is satisfied that they appear to meet the Contractor's Requirements.

The entire matter of design liability will be dealt with in more detail in Chapter 4 of this book; however, it should be noted at this time that the sixth recital does not require the contractor to check that the sub-contractor's proposals *do* meet the contractor's requirements but merely that 'they appear to meet' them.

2.3.7 Seventh recital

This recital relates to the provision by the sub-contractor of a priced schedule of activities in which each activity is priced so that the sum of those individual prices equals the sub-contract sum, excluding provisional sums and excluding the value of work for which approximate quantities are included in any bills of quantities.

This recital should be deleted if either article 3B is to apply (i.e. where the sub-contract works are to be completely re-measured and valued on a re-measurement basis) or where a priced activity schedule is not provided.

2.3.8 Eighth recital

The eighth recital states that the sub-contractor has been given a copy of, or a reasonable opportunity to inspect, any other documents and information relating to the provisions of the main contract in so far as they relate to the sub-contract works and which are listed in item 1.9 of the schedule of information.

It is essential that both the contractor and the sub-contractor ensure that this requirement is properly fulfilled, and that item 1.9 of the schedule of information correctly lists *only* the other documents and information which relate to the sub-contract works and which the sub-contractor has been given a copy of or a reasonable opportunity to inspect.

Given its importance, a sub-contractor should resist the inclusion of such a list of documents and information in the sub-contract agreement until such time that copies are provided to him or her or until he or she has been given a reasonable opportunity to inspect the documents and information. Obviously, once this occurs, the sub-contractor should then read and familiarise himself or herself with same.

2.3.9   Ninth recital

This recital says that where so stated in the sub-contract particulars (under item 1), the sub-contract is supplemented by the Framework Agreement(s) identified in those particulars.

2.3.10   Tenth recital

The tenth recital says that the supplemental sub-contract provisions as identified in the sub-contract particulars item 1 apply.

2.4   The articles

The articles set out certain basic elements of the sub-contract (e.g. the sub-contract documents, the sub-contract conditions, the sub-contract sum [or the sub-contract tender sum], the final sub-contract sum and the dispute resolution procedures).
Under the SBCSub/D/A, there are six articles as follows.

2.4.1   Article 1

This article lists the documents comprising the sub-contract documents:

- The agreement (i.e. SBCSub/D/A)
- The sub-contract particulars
- The schedule of information, plus all documents referenced within it and annexed to it
- The sub-contract conditions (i.e. SBCSub/D/C), incorporating thereto the JCT amendments and any schedule of modifications stated in the sub-contract particulars (item 1), and listed (and annexed) as a numbered document
- The numbered documents annexed to the agreement

2.4.2   Article 2

Article 2 states the basic obligation of the sub-contractor, which is to carry out and complete the sub-contract works (including the design as appropriate) in accordance with the sub-contract. Where the sub-contractor's obligations include the requirement to complete the design for the Sub-contractor's Designed Portion, this is to be in accordance with the CDM Regulations, and in line with such directions as the contractor may give for the integration of the design for the Sub-contractor's Designed Portion with the design for the main contract works as a whole.

2.4.3   Article 3A (sub-contract sum and final sub-contract sum) and Article 3B (sub-contract tender sum and final sub-contract sum)

Either article 3A or article 3B is completed as appropriate. In both cases, a sum of money in both words and figures needs to be inserted. The article that is not completed is to be deleted.

Article 3A is used where a sub-contract sum is applicable (i.e. where the works are valued on an adjustment basis, which means that the sub-contract sum only changes if the sub-contract so provides – for example, when variations are carried out).

Article 3B is used where a sub-contract tender sum is applicable (i.e. where the works are to be completely re-measured, usually referred to as the re-measurement basis).

2.4.4 Article 4

This article confirms that if any dispute or difference arises under the sub-contract, either party may refer it to adjudication in accordance with clause 8.2 of the sub-contract conditions.

2.4.5 Article 5

Article 5 relates to the position where arbitration is to be used to finally resolve any dispute or difference between the parties.

For arbitration to apply, the sub-contract particulars must be completed to record that article 5 and clauses 8.3–8.8 apply (i.e. by deleting the words 'do not apply'). If neither entry is deleted, the default position is that legal proceedings will apply and arbitration will not apply.

2.4.6 Article 6

When article 6 applies, the English courts are to have jurisdiction over any dispute or difference between the parties which arises out of or in connection with the sub-contract. As stated under article 5 above, unless the relevant entry is made within the sub-contract to select arbitration, the default position is that legal proceedings will apply and not arbitration.

The footnote for article 6 states that if the parties wish any dispute or difference to be determined by the courts of another jurisdiction (i.e. not the English courts), the appropriate amendment should be made to article 6.

2.5 The sub-contract particulars

The sub-contract particulars set out certain basic terms of the sub-contract.

In many cases, a default position has been provided in respect of the sub-contract particulars, and in such cases, an entry only need be made where it is intended that the default position is not to apply.

Where an entry is made, and this entry requires a continuation sheet, then such continuation sheets should be identified as such, signed or initialled by or on behalf of each party, and then annexed to the applicable agreement.

2.5.1   Sub-contract particulars number 1 – the conditions

This relates to the conditions to apply, and confirms that those conditions are to be as given in the SBCSub/D/C. The JCT amendments to the standard conditions are either those specifically noted or, if not noted, then the default position is that those JCT amendments which are current at the sub-contract base date (see section 2.5.3) will apply.

In addition, there is the facility for a schedule of modifications to the standard conditions to be incorporated. If no such schedule applies, the reference to a schedule of modifications should be deleted; however, if a schedule of modifications is provided, then it should be listed (and annexed) as a numbered document.

The schedule of modifications can obviously considerably alter the entire basis of the standard conditions, and such schedules are usually drafted to reflect amendments to the main contract and/or in an attempt to give an advantage to the contractor (and a disadvantage to the sub-contractor) in respect of particular clauses in the standard conditions printed text.

A sub-contractor must take careful note of the implications of the proposed schedule of modifications before agreeing to accept the said schedule as each proposed modification could have a serious implication upon the sub-contractor's contractual position.

Also, details of any applicable Framework Agreements should be inserted, and any applicable listed supplemental sub-contract provisions, as listed below, should be indicated.

Sub-contractor's particular number 1 operates such that the default position is that the listed sub-contract provisions apply unless, against each provision, the word 'applies' is struck through and the words 'does not apply' are left in place.

The listed supplemental sub-contract provisions (which are dealt with under Schedule 1 to the SBCSub/D/C) are:

- Collaborative working
- Health and safety
- Cost savings and value improvements
- Sustainable developments and environmental considerations
- Performance indicators and monitoring
- Notification and negotiation of disputes

In respect of the latter provision, both the contractor and the sub-contractor are to name their respective nominees to take part in the notification and negotiation of disputes process.

2.5.2   Sub-contract particulars number 2 – arbitration

Under this item, the parties agree whether arbitration will or will not apply.

If the parties agree that disputes and differences are to be determined by arbitration, then this entry must be completed to state that article 5 and clauses 8.3–8.8 will apply; if it is decided that arbitration is not to apply, then legal proceedings will apply.

If it is not decided by the parties that arbitration is or is not to apply, then the default position is that legal proceedings will apply.

2.5.3 Sub-contract particulars number 3 – base date

The sub-contract base date is to be inserted here. The sub-contract base date will not necessarily be the same as the main contract base date (although it often is). Usually, the sub-contract base date is a date that is close to the sub-contractor's tender or commencement date.

It should be noted that the sub-contract base date is not an arbitrary or irrelevant date; it is relevant, amongst other things, to:

- the standard sub-contract amendments that may apply,
- matters relating to any divergences from statutory requirements,
- the applicable standard method of measurement to be used, and
- the applicable definitions of the prime cost of daywork to be used.

2.5.4 Sub-contract particulars number 4 – address for notices

If there are particular addresses to be used for the service of written notices, they should be set out here. If no addresses are inserted, then the addresses that apply are those set out at the head of the agreement.

2.5.5 Sub-contract particulars number 5 – programme

Item 5.1 – preparation of sub-contractor's drawings

A time period in weeks is inserted here for the preparation of all necessary sub-contractor's drawings, etc. (co-ordination, installation, shop or builder's work or other drawings as appropriate), from receipt of the instruction to proceed with such preparation *and* from receipt of all other relevant drawings and specifications, etc. to the submission to the contractor of the drawings, etc. for comment.

Because the time period commences from:

- the receipt of the instruction to proceed with such preparation, and
- from receipt of all other relevant drawings and specifications,

the contractor must ensure that both the required instruction and all relevant drawings and specifications are provided in sufficient time to meet the programme requirements.

Item 5.2 – contractor's initial comments upon the sub-contractor's drawings

Under this item a time period in weeks is to be inserted for the contractor's initial comments upon the drawings, etc. (provided by the sub-contractor in line with item 5.1), from when the drawings, etc. have been received by the contractor to when the drawings, etc. are returned to the sub-contractor.

Although not stated, the implication is that the time period would apply in respect of individual drawings, etc. rather than apply to the date of receipt by the contractor of a set of drawings.

Note that the periods entered under items 5.1 and 5.2 need to take into account clause 2.6.2, which:

- requires the sub-contractor to provide to the contractor design documents and other information to enable the contractor to observe and perform his or her obligations relating to the main contract/the contractor's design submission procedure, and
- prohibits the sub-contractor from commencing any work in respect of a design document or other information which is subject to the contractor's design submission procedure before that procedure has been complied with.

Item 5.3 – pre-site commencement: procurement of materials, fabrication, delivery, etc.

A time period in weeks is to be inserted here for the procurement of materials, fabrication (where appropriate) and delivery to site prior to commencing work on site or work on site in each section.

The period inserted here would need to take into account any extended manufacturing/delivery times for any non-standard materials. It is conceivable that different periods may be required to be inserted for different sections (although it should be noted that the entry makes no specific provision for this possibility).

Consecutive total to sub-contract particular items 5.1–5.3

It is important to note that the periods indicated under items 5.1–5.3 are consecutive periods. This means that each individual time period inserted against these items should then be added together and inserted as a total period (for items 5.1–5.3) in the paragraph located between items 5.3 and 5.4 of the sub-contract particulars.

Item 5.4 – notice to commence

A time period in weeks is to be inserted for the period of notice required to commence works on site to enable a start to be made to the sub-contract works or to each section (if applicable).

This said time period (or periods) inserted under this item is not stated as being consecutive to the periods inserted for items 5.1–5.3 and, therefore, must be taken as being concurrent with those periods.

This position is made clear at item 5.5, which notes that the period (or periods) required for the carrying out of the sub-contract works on site commences after delivery of materials (i.e. the summated period of items 5.1–5.3, if applicable) *and* after the expiry of the period of notice to commence works (as item 5.4). The position is also confirmed under clause 2.3, which says that the on-site construction periods are subject 'to receipt by the Sub-contractor of notice to commence work'.

Therefore, by way of an example:

- A five-week notice period is stated under item 5.4.
- The total consecutive time period for items 5.1–5.3 is 11 weeks.
- If, however, the notice under item 5.5 was not given until week 10 of the 11-week period (for items 5.1–5.3), then even if the materials were delivered to site after the 11-week period, because of the *late* notice given under item 5.5, the five-week notice period would run from week 10. Accordingly, and in this example, the sub-contractor could conceivably not commence works until the expiry of a 15-week period (i.e. week 10 plus the five-week notice period required).

Despite this, it must be noted that, although there is nothing to prevent a sub-contractor from commencing prior to the stated notice period, and despite the fact that this may frequently occur, the sub-contractor is not contractually obliged to commence prior to the stated notice period.

The above issues highlight that the late issue of the notice required by item 5.4 could have a major impact upon the programme for the main contract works. The contractor must ensure he or she provides this notice at a time to suit the required start date on site for the whole or a section of the sub-contract works.

It should be noted that the notice required under sub-contract particulars item 5.4 relates only to the period of notice required to commence works on site to enable a start to be made to the sub-contract works or to each section; it does not relate to any off-site works or design works.

Item 5.5 – period required for the carrying out of the sub-contract works on site

Under this item, the period required for the carrying out of the sub-contract works on site is inserted, and this will run from whichever is the latter of the following:

- the delivery of materials (i.e. either the summated period of items 5.1–5.3 or the period of item 5.1 alone) or
- the expiry of the period of notice to commence works (as item 5.4)

The period or periods stated are the total 'on-site' working periods. It may be that this total period is not carried out in one consecutive period.

If sectional completion does not apply, then only one period (in weeks) needs to be inserted. However, if sectional completion does apply, then for each section, a period (in weeks) needs to be inserted. If there is insufficient room on the entries provided in the sub-contract agreement for every required section, further sheets should be used and annexed to the sub-contract agreement as appropriate.

When agreeing to a period, the sub-contractor needs to take into consideration the information provided under item 3 of the schedule of information, which relates to site opening and closing times and periods.

Item 5.6 – further details or arrangements that may qualify or clarify the period required for the carrying out of the sub-contract works on site

Obviously, the sub-contract works are not usually undertaken in isolation. There will clearly normally be other trades, obligations upon the contractor imposed by the main contract in respect of the main contract works (or sections), access and co-ordination issues that may affect the sub-contract works. Accordingly, the contractor will often seek to impose certain obligations on the sub-contractor (working to the contractor's programme, integration with other trades, sequence, carrying out the works in several visits, etc.) to manage and meet his or her obligations under the main contract.

In terms of progress of the sub-contract works in relation to the main contract works, the contractor has some protection by clause 2.3, which requires the sub-contractor to carry out and complete his or her work 'reasonably in accordance with the progress of the Main Contract works or each relevant Section of them'. However, it is always safer to rely on some specific requirement, rather than a more generalised obligation. This being the more so because the generalised obligation being relied upon in a similar clause in the DOM/1 form of the sub-contract was considered by Judge Gilliland[4] as *not* requiring:

> 'The sub-contractor to plan his sub-contract work so as to fit in with either any scheme of work of the main contractor or to finish any part of the sub-contract works by a particular date so as to enable the main contractor to proceed with other parts of the work'.

Accordingly, under item 5.6, any further details or arrangements that are relevant to the carrying out of the sub-contract works on site (or a section thereof) should be inserted.

If there is insufficient room on the printed form itself, further sheets should be used and should be annexed to the printed form.

Some limited examples being:

- any particular programme requirement or constraints;
- long lead in time for procurement of materials;
- specific sequence of working required (e.g. first floor, third floor, then basement);
- activities that must be completed by a certain date to accord with a section under the main contract;
- access restrictions;
- return visits;
- off-site periods;

---

[4]*Piggott Foundations Ltd* v. *Shepherd Construction Ltd* (1993) 67 BLR 53.

- key milestones/specific dates to be met (i.e. completion of first fix; commissioning, etc.); and/or
- integration and interdependency of sub-contract works with other sub-contractors, etc.

It is important that all entries are carefully and accurately completed, not least because such entries may have a major impact upon some or all of the periods in the preceding entries, particularly the period for completion required by the sub-contractor.

2.5.6   Sub-contract particulars number 6 – attendances

The entries for attendance are split into four sections as follows:

Item 6.1 – items of attendance which the contractor will provide to the sub-contractor free of charge

A basic list of possible attendance items to be provided free of charge by the contractor is set out in the printed form, which is to be amended and/or added to (as appropriate). If there is insufficient room, further sheets should be used and annexed to the form.

It is important that this list is carefully considered and correctly amended because all attendances (*other than those finally listed*) are to be provided by the sub-contractor.

Item 6.2 – Joint Fire Code: additional items

The main contract particulars included in the schedule of information should state whether or not the Joint Fire Code applies and, if it does, whether or not the requirements in respect of Large Projects apply. The Joint Fire Code is published by the Fire Protection Association,[5] and it provides guidance on fire safety on site as well as the prevention and detection of fire.

If the Joint Fire Code does apply, it will be the Joint Fire Code that was current as at the main contract base date (not the sub-contract base date).

If the Joint Fire Code does apply, the sub-contractor is expected to comply with that code, and the contractor is to provide the attendances listed under item 6.2.

If the requirements in respect of Large Projects are applicable to the Joint Fire Code, the contractor must also provide the service of an appropriate number of fire marshals.

Item 6.3 – location of sub-contractor's temporary buildings

Item 6.3 does not apply (and should be deleted) if:

- the sub-contractor is not to provide temporary buildings at all, or
- if the sub-contractor is to provide temporary buildings, but these are to be located more than 6 m from the building under construction.

---

[5]The Fire Protection Association (FPA) is the UK's national fire safety organisation.

If the Joint Fire Code applies to the project and the sub-contractor is to provide temporary buildings either within the building under construction and/or within 6 m of the building under construction, then:

■ the sub-contractor is to ensure that the temporary buildings are constructed in compliance with the requirements of paragraphs 12.4 and 12.8 of the Joint Fire Code, and

■ if the project is a Large Project under the Joint Fire Code, then the contractor is to connect (if he or she is responsible for the connection) an installed fire detection system to a central alarm-receiving station.

Item 6.4 – clearance of rubbish resulting from the carrying out of the sub-contract works

If all rubbish resulting from the carrying out of the sub-contract works is to be disposed of off site (by the sub-contractor), then no entry needs to be made.

However, if there are specific requirements for the manner in which rubbish is to be disposed of, then details should be inserted under this item; for example, the sub-contractor may be required to dispose of his or her rubbish to a central skip provided by the contractor.

2.5.7   Sub-contract particulars number 7 – interim payments

Under this item the first payment due date is inserted. If no date is entered (and unless clause 4.9.2 applies), the first payment due date shall (unless otherwise agreed in writing) be the date 1 month after the date of commencement of the sub-contract works on site.

Also, it needs to be noted if clause 4.9.2 is not marked to apply or not to apply, then the default position is that clause 4.9.2 *does not apply*.

2.5.8   Sub-contract particulars number 8 – listed items

If the contractor and the sub-contractor have agreed that certain goods and materials will be paid for prior to delivery to site (i.e. off site), these should be listed under this sub-contract particulars item. Payment for such listed items is subject to the preconditions of clause 4.14 having been fulfilled.

If a bond is required in respect of these items, this should be stated and the maximum financial liability of the surety should also be stated. The form of the bond is set out at Schedule 3 Part 1 of the SBCSub/D/C.

2.5.9   Sub-contract particulars number 9 – retention

An entry only needs to be made here if the retention percentage is to be different to the default percentage of 3%.

The retention percentage can be at any level that the parties agree, and it does not have to be the same percentage as that applies to the main contract.

There is a facility for a minimum retention amount to be specified, which would be £250.00 unless a greater amount is stated.

There is also the opportunity for a retention release date (for the final release of retention) to be inserted. If no date is inserted, the retention release date would be the day after the expiry of a period from the date for completion of the main contract works (or the last section of them) equivalent to the rectification period for the main contract works or a section thereof *plus a further 6 months.* The reason for a retention release date needing to be provided, and for a default provision applying, is because the final release of the sub-contractor's retention cannot be dependent upon the non-issue of a certificate under the main contract (this is dealt with in more detail in Chapter 8 of this book). Because of this default provision, there is a provision under this sub-contract particulars number 9 for the date for completion of the main contract works (or the last section thereof) and the rectification period for the main contract works (or the last section thereof) to be inserted.

2.5.10   Sub-contract particulars number 10 – retention bond

An entry only needs to be made here if a retention bond is required in lieu of retention monies being held. The form of the retention bond is to be as contained under Schedule 3 Part 2 of the SBCSub/D/C (dealt with later with in this book in Chapter 8).

The default position is that the retention bond will not apply, and it will also not apply unless the relevant particulars are provided under this sub-contract particulars item. The particulars to be provided being the maximum aggregate sum for the purposes of clause 2 of the retention bond and the expiry date for the purposes of clause 6.3 of the retention bond.

It should be noted that a retention bond may be provided by a sub-contractor irrespective of whether or not a retention bond is provided by the main contractor under the terms of the main contract.

2.5.11   Sub-contract particulars number 11 – fluctuations

The fluctuation options available are options A, B and C.[6] These options are set out in Schedule 4 of the SBCSub/D/C, which is dealt with later within this book in Chapter 8.

The only restriction regarding the choice of fluctuations made is that, where option A or B applies to the main contract, it is a requirement of that contract that the same option applies to any sub-contract.

However, where option C applies to the main contract, the parties are free to agree that any of the options A, B or C applies to the sub-contract.

Where option A or B applies, a percentage addition to the fluctuation payments or allowances applicable is to be inserted. Where option B applies, a list of basic transport charges is to be provided. This list may either be provided

---

[6]Under the ICSub and the ICSub/D, only fluctuation option A applies.

on the SBCSub/D/A or be contained on a separate document that is simply referred to in the appropriate part of the SBCSub/D/A. Of course, if a separate list is provided, it should be included as a numbered document.

2.5.12   Sub-contract particulars number 12 – dayworks

This sub-contract particular item is used if the numbered documents do not include a schedule of daywork rates or prices (for labour, materials and plant), which is to be used to value daywork items. In such circumstances, daywork is to be calculated in accordance with the definition or definitions identified under this sub-contract particulars item, together with the percentage entered for each section of the prime cost.

The three definitions referred to in this sub-contract particulars item (one or more of which may be chosen) are those agreed between the Royal Institution of Chartered Surveyors and:

- the Construction Confederation,
- the Electrical Contractors Association, and
- the Heating and Ventilating Contractors' Association.

The three categories of prime cost are labour, materials and plant:

- Labour: In respect of labour (which encompasses tradesmen, craftsmen and non-skilled labour), the options are either to:
  - ○ use one or more of the three definitions noted above to obtain the labour daywork base rate at the time that the daywork was carried out, and then apply the appropriate percentage addition for labour as inserted against the appropriate definition;
    or
  - ○ use the all-in rate inserted in sub-contract particulars number 12 in lieu of the cost of labour calculated in accordance with one of the definitions.
- Materials: For materials, the only option is to apply the appropriate percentage addition as inserted onto the prime cost of materials.
- Plant: In respect of plant, the only option is to use one or more of the three definitions noted above to obtain the plant daywork base rate at the time that the daywork was carried out, and then apply the appropriate percentage addition onto the schedule of plant hire rates (if appropriate) as inserted against the appropriate definition.

2.5.13   Sub-contract particulars number 13 – insurance: personal injury and property damage

The insurance cover entered here is the minimum level required in respect of death or personal injury (other than to employees covered under the sub-contractors' statutory employers' liability policy) and in respect of damage to property (excluding specified perils' damage to the main contract works and site materials) up to the terminal date.

The terminal date and the specified perils are defined under clause 6.1 of the sub-contract conditions.

The level of cover required should not normally exceed the level stated in the main contract particulars. However, the level could be less by agreement if the sub-contractor is unable to provide the level of cover required, and if the contractor's insurance covers for any shortfall on cover.

2.5.14 Sub-contract particulars number 14 – incorporation of the sub-contract works into the main contract works

The insertions here are to be for those elements of the sub-contract works that the contractor is prepared to regard as fully, finally and properly incorporated into the main contract works prior to practical completion of the sub-contract works or a section, as applicable, and the extent to which each of the listed elements needs to be carried out to achieve the status of being fully, finally and properly incorporated into the main contract works. If there is insufficient space to make the required entries, then a separate sheet may be used, which should then be annexed to the sub-contract particulars.

It should be noted that once incorporated, clause 6.7.2 provides important protection to the sub-contractor in terms of loss or damage to same as follows:

'Where, during the progress of the Sub-contract works sub-contract materials or goods have been fully, finally and properly incorporated into the Main Contract works before the Terminal Date, then, in respect of loss or damage to any of the Sub-contract works so incorporated that is caused by the occurrence of a peril other than a Specified Peril, the sub-contractor shall only be responsible for the cost of restoration of such work lost or damaged and removal and disposal of any debris in accordance with clause 6.7.3 to the extent that such loss or damage is caused by the negligence, breach of statutory duty, omission or default of the sub-contractor or of any of the sub-contractors' Persons'.

The terminal date is defined under clause 6.1 of the sub-contract conditions, and for the purposes of this matter, ordinarily, it will be the date of practical completion of the sub-contract works, or if applicable, a section thereof.

Obviously, the entries under this sub-contract particulars item must be carefully and precisely set out in order that the contractor and the sub-contractor can readily determine when certain elements of the sub-contract works have been fully, finally and properly incorporated into the main contract works prior to the terminal date.

2.5.15 Sub-contract particulars number 15 – sub-contractor's designed works professional indemnity insurance

The entry here relates to the level of cover required in respect of professional indemnity insurance in respect of clause 6.10 of the SBCSub/D/C.

It should be particularly noted that if no amount of cover is inserted, clause 6.10 will not apply regardless of whether or not the other parts of this item are completed.

The amount of cover required either:

- relates to claims or a series of claims arising out of one event, or
- is the aggregate amount for any one period of insurance. Unless stated otherwise, a period of insurance for these purposes is 1 year.

One or other of the above options is to be deleted.

If a level of cover that is different to the full indemnity cover noted above is required for pollution/contamination claims, this level of cover should be entered. If no level of cover is stated, then the required level of cover will be the full amount of the indemnity cover stated above.

The period (in terms of years) that the professional indemnity insurance needs to be maintained from the date of practical completion of the sub-contract works needs to be entered. The choices are 6 years, 12 years or some other period in years. If no entry is made, the expiry date will be taken as being 6 years from the date of practical completion of the main contract works (not from the practical completion of the sub-contract works).

As a general guide, if the sub-contract is executed under hand, a professional indemnity insurance period of 6 years after the date of practical completion of the sub-contract works would be adequate (because the limitation period for instituting proceedings is 6 years where a contract is executed under hand); but if the sub-contract is executed as a deed, a professional indemnity insurance period of 12 years after the date of practical completion of the sub-contract works would be required (because the limitation period for instituting proceedings is extended to 12 years where a contract is executed as a deed).

2.5.16   Sub-contract particulars number 16 – settlement of disputes

Relevant entries are made here in respect of adjudication, and if it is to apply, arbitration.

The entries are self-explanatory. However, in respect of adjudication, an adjudicator may be named and/or a nominator of an adjudicator may be selected.

If an adjudicator is named, that named adjudicator should be first approached when a party requires a reference to adjudication to be made.

Even if an adjudicator is named, it is recommended that a nominator of an adjudicator is selected in case the named adjudicator is unwilling or unable to act for some reason.

2.5.17   Sub-contract particulars number 17 – numbered documents

Any numbered documents (and, as appropriate, any annexures to any numbered documents) should be listed under this sub-contract particulars item.

The said documents should be numbered sequentially, should be initialled or signed by each party and should be annexed to the sub-contract particulars.

The numbered documents should comprise all the documents which are to be the sub-contract documents other than those already listed at article 1, namely the sub-contract agreement, the sub-contract conditions, and the schedule of information and its listed annexures.

2.6 Attestation forms

Attestation simply means to certify the validity of the sub-contract agreement, and this validity is confirmed by the parties executing the sub-contract agreement in one of two ways:

- Under hand
- As a deed

2.6.1 Execution under hand or as a deed

If a sub-contract is executed under hand, the limitation period for commencing proceedings due to a breach of the sub-contract is 6 years, whereas if it is executed as a deed, the limitation period is extended to 12 years.

The question of the limitation period often becomes an issue when latent defects become apparent. A latent defect is a defect that for some reason or another is concealed or lies dormant, as distinct from a patent defect, which is open and obviously exists.

Normally, a contractor would wish to have the same limitation period for any recourse against a sub-contractor as exists against the contractor from the employer. Therefore, if the main contract was executed as a deed (as is usually the case), the contractor would normally require the sub-contract to be executed as a deed. However, if the main contract was executed under hand, the contractor would normally accept that the sub-contract should be executed under hand.

Where the sub-contract is to be executed under hand, the sub-contract contains guidance notes on how this execution should be carried out. Whilst the entries are relatively self-explanatory, they should nevertheless be meticulously followed and re-checked by the parties. All too often, such important matters are not given the careful attention they deserve.

If executed as a deed, the parties may execute the agreement either as a company (or other body corporate) or as an individual, as appropriate. In all cases, the contractor's name and the sub-contractor's name should be inserted in the appropriate place on the page headed 'Execution as a Deed'.

Again, where this option is chosen, the sub-contract contains detailed notes on executing the sub-contract as a deed. Whilst the entries are relatively self-explanatory, as stated above, they must be meticulously followed and re-checked by the parties.

Where foreign companies are involved, they can execute deeds (under the Companies Act 1989, as applied by the Foreign Companies [Execution of Documents] Regulations 1994 and the 1995 amendments to those regulations).

However, because of the complications that may arise when foreign companies are involved, it is often best to obtain professional advice regarding the proposed execution method.

In cases where the forms of attestation set out in the sub-contracts are not appropriate (e.g. in the case of certain housing associations and partnerships), then the appropriate forms may be inserted by the parties in lieu of the forms provided. Again, professional advice should be sought if this is required.

2.6.2   Why do we have limitation periods?

The principle behind limitation periods is to provide a cut-off date after which no claim can be brought (i.e. the claim becomes time barred). The reason behind this is that the law regards it as undesirable that liability could exist without time limit. Limitation periods are set down for both contract and tort in the Limitation Act 1980.

If a contractor's action is in contract against a sub-contractor in respect of defects, the position may depend upon whether the sub-contract contains an express clause whereby the sub-contractor indemnifies the contractor against breaches of contract. Such an indemnity is provided under clause 2.5 of the SBCSub/D/C, and the contractor's action upon the indemnity clause may not be statute barred until the expiry of the limitation period (i.e. 6 or 12 years, as appropriate) between the employer and the contractor under the main contract.[7]

2.6.3   Does the Latent Damage Act 1986 affect the sub-contract's limitation period?

The Latent Damage Act 1986 does not affect the existing law on limitation in contract cases[8] where it has been found that section 14A of the Limitation Act 1980 could not be applied to actions in contract.

One particular area of note is that under section 32 of the Limitation Act 1980, the limitation period can be extended if the contractor (or the sub-contractor) has deliberately concealed the defect, and this may apply even where the employer (or the contractor, as appropriate) had the benefit of agents overseeing the works.[9]

However, simply proceeding with the works does not necessarily give rise to a deliberate concealment; to establish deliberate concealment under section 32 of the Limitation Act 1980, the claimant must show that the defendant:

1. has taken active steps to conceal a breach of duty after he or she has become aware of it, or
2. is guilty of deliberate wrongdoing and has concealed or has failed to disclose such wrongdoing in circumstances where it is unlikely to be discovered for some time.

---

[7]*County and District Properties* v. *Jenner* (1976) 3 BLR 38.

[8]*Iron Trades Mutual Insurance Co Ltd* v. *JK Buckenham Ltd* [1990] 1 All ER 808.

[9]*Lewisham Borough Council* v. *Leslie & Co* (1979) 12 BLR 22, CA.

2.7   Schedule of information

It is important that the schedule of information is accurately provided by the contractor and noted by the sub-contractor, particularly because clause 2.5.1 states:

'Insofar as the Contractor's obligations under the Main Contract, as identified in or by the Schedule of information, relate and apply to the Sub-contract works (or any part of them), the Sub-contractor shall observe, perform and comply with those obligations, including (without limitation) those under clauses 2.10 (Levels and setting out), 2.21 (Fees or charges legally demandable), 2.22 and 2.23 (Royalties and patent rights) and 3.22 and 3.23 (antiquities) of the Main Contract Conditions, and shall indemnify and hold harmless the Contractor against and from:

1. any breach, non-observance or non-performance by the Sub-contractor or his employees or agents of any of the provisions of the Main Contract; and
2. any act or omission of the Sub-contractor or his employees or agents which involves the Contractor in any liability to the Employer under the provisions of the Main Contract'.

2.7.1   Item 1

Item 1 of the schedule of information provides the following information (in respect of the main contract):

- The name of the:
  - employer
  - the architect/contract administrator[10]
  - the Principal Designer (if named in the main contract)
- Confirmation that the Principal Contractor is the contractor, or if this is not the case, the name of the Principal Contractor is inserted here.
- Identification of the main contract, plus (where relevant):
  - any applicable JCT amendments to it (i.e. published by the JCT)
  - details of any schedule of modifications, that is, any employer (non-standard) amendments to the main contract's published text. Where applicable, the schedule of modifications should be annexed to the sub-contract agreement.
- A copy of the main contract particulars must be annexed to the sub-contract agreement. This is directly copied from the main contract between the employer and the contractor.
- Any changes to the main contract particulars. For example, any change to the completion date for the main contract works or of any section(s).

---

[10]The Employer's Agent under the DBSub/A.

- Confirmation as to whether or not the main contract works is divided into sections, and if applicable, details of those sections (if the information is not already provided in the main contract particulars).
- Finally, under item 1.9 of the schedule of information, any other documents and information relating to the provisions of the main contract of which the sub-contractor has had a copy or a reasonable opportunity to inspect.
- Any documents annexed to the schedule of information should be signed or initialled by or on behalf of each party.

2.7.2 Item 2

Item 2 of the schedule of information notes that a copy of the Construction Phase Plan relevant to the Sub-contract Works (together with any developments to the Construction Phase Plan by the Principal Contractor notified to the sub-contractor before or during the progress of the sub-contract works), should be annexed to the schedule of information, and should be signed or initialled by or on behalf of each party.

2.7.3 Item 3

Item 3 of the schedule of information sets out certain programme information, particularly in respect of the opening and closing times on site, and also the earliest and latest starting date for the sub-contract works to be carried out on site.

2.8 Supplementary particulars

The supplementary particulars contains information that is needed for the formula adjustment for fluctuations, and these particulars only need to be given when fluctuation option C applies.

2.9 Equivalent sub-contract provisions

In the table below, the equivalent sub-contract provisions to that within the SBCSub/D/A or SBCSub/D/C in the text above are listed for the SBCSub/A or SBCSub/C, DBSub/A or DBSub/C, ICSub/A or ICSub/C and ICSub/D/A or ICSub/D/C.

SBCSub/D/A or SBCSub/D/C	SBCSub/A or SBCSub/C	DBSub/A or DBSub/C	ICSub/A or ICSub/C	ICSub/D/A or ICSub/D/C
First Recital	First Recital	First Recital	First Recital	First Recital
Second Recital	Second Recital	Second Recital	Second Recital	Second Recital
Third Recital	Not applicable to this Sub-contract as this Recital relates to Sub-contractor's design, which is not provided for under this Sub-contract	Third Recital	Not applicable to this Sub-contract as this Recital relates to Sub-contractor's design, which is not provided for under this Sub-contract	Third Recital
Fourth Recital	Not applicable to this Sub-contract as this Recital relates to Sub-contractor's design, which is not provided for under this Sub-contract	Fourth Recital	Not applicable to this Sub-contract as this Recital relates to Sub-contractor's design, which is not provided for under this Sub-contract	Fourth Recital
Fifth Recital	Not applicable to this Sub-contract as this Recital relates to Sub-contractor's design, which is not provided for under this Sub-contract	Fifth Recital	Not applicable to this Sub-contract as this Recital relates to Sub-contractor's design, which is not provided for under this Sub-contract	Fifth Recital
Sixth Recital	Not applicable to this Sub-contract as this Recital relates to Sub-contractor's design, which is not provided for under this Sub-contract	Sixth Recital	Not applicable to this Sub-contract as this Recital relates to Sub-contractor's design, which is not provided for under this Sub-contract	Sixth Recital
Seventh Recital	Third Recital	Seventh Recital	Third Recital	Seventh Recital
Eighth Recital	Fourth Recital	Eighth Recital	Fourth Recital	Eighth Recital
Ninth Recital	Fifth Recital	Ninth Recital	Fifth Recital	Ninth Recital
Tenth Recital	Sixth Recital	Tenth Recital	Sixth Recital	Tenth Recital
Article 1	Article 1	Article 1	Article 1	Article 1
Article 2	Article 2	Article 2	Article 2	Article 2
Article 3A	Article 3A	Article 3A	Article 3A	Article 3A
Article 3B	Article 3B	Article 3B	Article 3B	Article 3B
Article 4	Article 4	Article 4	Article 4	Article 4
Article 5	Article 5	Article 5	Article 5	Article 5
Article 6	Article 6	Article 6	Article 6	Article 6

(Continued)

SBCSub/D/A or SBCSub/D/C	SBCSub/A or SBCSub/C	DBSub/A or DBSub/C	ICSub/A or ICSub/C	ICSub/D/A or ICSub/D/C
Sub-contract Particulars Item 1	Sub-contract Particulars Item 1	Sub-contract Particulars Item 1	Sub-contract Particulars Item 1	Sub-contract Particulars Item 1
Sub-contract Particulars Item 2	Sub-contract Particulars Item 2	Sub-contract Particulars Item 2	Sub-contract Particulars Item 2	Sub-contract Particulars Item 2
Sub-contract Particulars Item 3	Sub-contract Particulars Item 3	Sub-contract Particulars Item 3	Sub-contract Particulars Item 3	Sub-contract Particulars Item 3
Sub-contract Particulars Item 4	Sub-contract Particulars Item 4	Sub-contract Particulars Item 4	Sub-contract Particulars Item 4	Sub-contract Particulars Item 4
Sub-contract Particulars Item 5	Sub-contract Particulars Item 5	Sub-contract Particulars Item 5	Sub-contract Particulars Item 5	Sub-contract Particulars Item 5
Sub-contract Particulars Item 5.1	Not applicable to this Sub-contract as this Sub-contract Particular relates to Sub-contractor's design, which is not provided for under this Sub-contract	Sub-contract Particulars Item 5.1	Not applicable to this Sub-contract as this Sub-contract Particular relates to Sub-contractor's design, which is not provided for under this Sub-contract	Sub-contract Particulars Item 5.1
Sub-contract Particulars Item 5.2	Not applicable to this Sub-contract as this Sub-contract Particular relates to Sub-contractor's design, which is not provided for under this Sub-contract	Sub-contract Particulars Item 5.2	Not applicable to this Sub-contract as this Sub-contract Particular relates to Sub-contractor's design, which is not provided for under this Sub-contract	Sub-contract Particulars Item 5.2
Sub-contract Particulars Item 5.3	Sub-contract Particulars Item 5.1	Sub-contract Particulars Item 5.3	Sub-contract Particulars Item 5.1	Sub-contract Particulars Item 5.3
Sub-contract Particulars Item 5.4	Sub-contract Particulars Item 5.2	Sub-contract Particulars Item 5.4	Sub-contract Particulars Item 5.2	Sub-contract Particulars Item 5.4
Sub-contract Particulars Item 5.5	Sub-contract Particulars Item 5.3	Sub-contract Particulars Item 5.5	Sub-contract Particulars Item 5.3	Sub-contract Particulars Item 5.5
Sub-contract Particulars Item 5.6	Sub-contract Particulars Item 5.4	Sub-contract Particulars Item 5.6	Sub-contract Particulars Item 5.4	Sub-contract Particulars Item 5.6

SBCSub/D/A or SBCSub/D/C	SBCSub/A or SBCSub/C	DBSub/A or DBSub/C	ICSub/A or ICSub/C	ICSub/D/A or ICSub/D/C
Sub-contract Particulars Item 6	Sub-contract Particulars Item 6	Sub-contract Particulars Item 6	Sub-contract Particulars Item 6	Sub-contract Particulars Item 6
Sub-contract Particulars Item 6.1	Sub-contract Particulars Item 6.1	Sub-contract Particulars Item 6.1	Sub-contract Particulars Item 6.1	Sub-contract Particulars Item 6.1
Sub-contract Particulars Item 6.2	Sub-contract Particulars Item 6.2	Sub-contract Particulars Item 6.2	Sub-contract Particulars Item 6.2	Sub-contract Particulars Item 6.2
Sub-contract Particulars Item 6.3	Sub-contract Particulars Item 6.3	Sub-contract Particulars Item 6.3	Sub-contract Particulars Item 6.3	Sub-contract Particulars Item 6.3
Sub-contract Particulars Item 6.4	Sub-contract Particulars Item 6.4	Sub-contract Particulars Item 6.4	Sub-contract Particulars Item 6.4	Sub-contract Particulars Item 6.4
Sub-contract Particulars Item 7	Sub-contract Particulars Item 7	Sub-contract Particulars Item 7	Sub-contract Particulars Item 7	Sub-contract Particulars Item 7
Sub-contract Particulars Item 8	Sub-contract Particulars Item 8	Sub-contract Particulars Item 8	Sub-contract Particulars Item 9	Sub-contract Particulars Item 9
Sub-contract Particulars Item 9	Sub-contract Particulars Item 9	Sub-contract Particulars Item 9	Sub-contract Particulars Item 8	Sub-contract Particulars Item 8
Sub-contract Particulars Item 10	Sub-contract Particulars Item 10	Sub-contract Particulars Item 10	Not applicable to this Sub-contract as this Sub-contract Particular relates to a Retention Bond, which is not provided for under this Sub-contract	Not applicable to this Sub-contract as this Sub-contract Particular relates to a Retention Bond, which is not provided for under this Sub-contract
Sub-contract Particulars Item 11	Sub-contract Particulars Item 11	Sub-contract Particulars Item 11	Sub-contract Particulars Item 10	Sub-contract Particulars Item 10
Sub-contract Particulars Item 12	Sub-contract Particulars Item 12	Sub-contract Particulars Item 12	Sub-contract Particulars Item 11	Sub-contract Particulars Item 11
Sub-contract Particulars Item 13	Sub-contract Particulars Item 13	Sub-contract Particulars Item 13	Sub-contract Particulars Item 12	Sub-contract Particulars Item 12
Sub-contract Particulars Item 14	Sub-contract Particulars Item 14	Sub-contract Particulars Item 14	Sub-contract Particulars Item 13	Sub-contract Particulars Item 13

(Continued)

SBCSub/D/A or SBCSub/D/C	SBCSub/A or SBCSub/C	DBSub/A or DBSub/C	ICSub/A or ICSub/C	ICSub/D/A or ICSub/D/C
Sub-contract Particulars Item 15	Not applicable to this Sub-contract as this Sub-contract Particular relates to Sub-contractor's design, which is not provided for under this Sub-contract	Sub-contract Particulars Item 15	Not applicable to this Sub-contract as this Sub-contract Particular relates to Sub-contractor's design, which is not provided for under this Sub-contract	Sub-contract Particulars Item 12
Sub-contract Particulars Item 16	Sub-contract Particulars Item 15	Sub-contract Particulars Item 16	Sub-contract Particulars Item 14	Sub-contract Particulars Item 14
Sub-contract Particulars Item 17	Sub-contract Particulars Item 16	Sub-contract Particulars Item 17	Sub-contract Particulars Item 15	Sub-contract Particulars Item 15
Sub-contractor's Designed Portion	Not applicable to this Sub-contract as Sub-contractor's design is not provided for under this Sub-contract	Sub-contractor's Designed Portion	Not applicable to this Sub-contract as Sub-contractor's design is not provided for under this Sub-contract	Sub-contractor's Designed Portion
Schedule of Information Item 1	Schedule of Information Item 1	Schedule of Information Item 1	Schedule of Information Item 1	Schedule of Information Item 1
Schedule of Information Item 1.9	Schedule of Information Item 1.9	Schedule of Information Item 1.9	Schedule of Information Item 1.9	Schedule of Information Item 1.9
Schedule of Information Item 2	Schedule of Information Item 2	Schedule of Information Item 2	Schedule of Information Item 2	Schedule of Information Item 2
Schedule of Information Item 3	Schedule of Information Item 3	Schedule of Information Item 3	Schedule of Information Item 3	Schedule of Information Item 3
Clause 2.3	Clause 2.3	Clause 2.3	Clause 2.2	Clause 2.2
Clause 2.5	Clause 2.5	Clause 2.5	Clause 2.4	Clause 2.4
Clause 2.6.2	Not applicable to this Sub-contract as this Clause relates to Sub-contractor's design, which is not provided for under this Sub-contract	Clause 2.6.2	Not applicable to this Sub-contract as this Clause relates to Sub-contractor's design, which is not provided for under this Sub-contract	Clause 2.5.3
Clause 4.14	Clause 4.14	Clause 4.14	Clause 4.11	Clause 4.11
Clause 6.1	Clause 6.1	Clause 6.1	Clause 6.1	Clause 6.1
Clause 6.7.2	Clause 6.7.2	Clause 6.7.2	Clause 6.7.2	Clause 6.7.2

SBCSub/D/A or SBCSub/D/C	SBCSub/A or SBCSub/C	DBSub/A or DBSub/C	ICSub/A or ICSub/C	ICSub/D/A or ICSub/D/C
Clause 6.10	Not applicable to this Sub-contract as this Clause relates to Sub-contractor's design, which is not provided for under this Sub-contract	Clause 6.10	Not applicable to this Sub-contract as this Clause relates to Sub-contractor's design, which is not provided for under this Sub-contract	Clause 6.14
Clause 6.11	Not applicable to this Sub-contract as this Clause relates to Sub-contractor's design, which is not provided for under this Sub-contract	Clause 6.11	Not applicable to this Sub-contract as this Clause relates to Sub-contractor's design, which is not provided for under this Sub-contract	Clause 6.15
Clause 8.2	Clause 8.2	Clause 8.2	Clause 8.2	Clause 8.2
Clause 8.3	Clause 8.3	Clause 8.3	Clause 8.3	Clause 8.3
Clause 8.4	Clause 8.4	Clause 8.4	Clause 8.4	Clause 8.4
Clause 8.5	Clause 8.5	Clause 8.5	Clause 8.5	Clause 8.5
Clause 8.6	Clause 8.6	Clause 8.6	Clause 8.6	Clause 8.6
Clause 8.7	Clause 8.7	Clause 8.7	Clause 8.7	Clause 8.7
Clause 8.8	Clause 8.8	Clause 8.8	Clause 8.8	Clause 8.8
Schedule 1	Schedule 1	Schedule 1	Schedule 3	Schedule 3
Schedule 3 Part 1	Schedule 3 Part 1	Schedule 3 Part 1	Schedule 1	Schedule 1
Schedule 3 Part 2	Schedule 3 Part 2	Schedule 3 Part 2	Not applicable to this Sub-contract as this Part of the Schedule relates to a Retention Bond, which is not provided for under this Sub-contract	Not applicable to this Sub-contract as this Part of the Schedule relates to a Retention Bond, which is not provided for under this Sub-contract
Schedule 4	Schedule 4	Schedule 4	Schedule 2	Schedule 2

3 Definitions and Interpretations

3.1 Introduction

The definitions and interpretations section of sub-contracts is often overlooked, even though that section contains some of the most important information in the sub-contracts.

Against this background, in this chapter, the following matters will be dealt with, using the sub-contract clause references, etc. contained in the SBCSub-/D/A or SBCSub/D/C (as appropriate). Whilst it is clearly beyond the scope of this book to review every nuance of the other sub-contract forms under consideration, the equivalent provisions (where applicable) within the SBCSub/A or SBCSub/C, the DBSub/A or DBSub/C, the ICSub/A or ICSub/C and the ICSub/D/A or ICSub/D/C are given at the table at the end of this chapter. It must be emphasised that before considering a particular issue, the actual terms of the appropriate edition of the relevant sub-contract should be reviewed by the reader (and/or legal advice should be sought as appropriate) before proceeding with any action/inaction in respect of the sub-contract in question.

In this chapter, the following matters are dealt with:

- Definitions
- Interpretation
- Reckoning periods of days
- Contracts (Right of Third Parties) Act 1999
- Service of notices and other documents
- Effect of the final payment notice
- Applicable law

3.2 Definitions

The definitions of many terms are provided under clause 1.1, which says that:

> 'Unless the context otherwise requires or this Sub-contract specifically provides otherwise, the following words and phrases, where they appear in capitalised form, in the Sub-contract Agreement or these Conditions, shall have the meanings stated or referred to below.'

The JCT 2011 Building Sub-contracts, First Edition. Peter Barnes and Matthew Davies.
© 2016 John Wiley & Sons, Ltd. Published 2016 by John Wiley & Sons, Ltd.

This clause is very important because, when dealing with the sub-contracts, any defined terms as set out under clause 1.1 must be stated in the capitalised form, rather than in the lowercase form, so that there is no doubt about what is actually being referred to.

Therefore, when corresponding about, for example, the sub-contract works, 'Sub-contract Works' should be used rather than the more generalised 'sub-contract works', to make it clear that the sub-contract works as defined in the sub-contract are being specifically referred to.

Despite the importance of the above definition clause, this book does not always and/or does not generally follow this convention and/or protocol. Therefore, very often, a word that is defined in the sub-contract is included in the text of this book in lower case – for example, 'Variation' may simply be included as 'variation'. The only exception to the above convention is where it is considered that this may cause unnecessary ambiguity to the reader.

With the above point in mind, the reader is strongly advised to refer to the various definitions as referred to above when reading the various chapters of this book.

3.3   Interpretation

3.3.1   Reference to clauses, etc.

Clause 1.2 simply clarifies that, unless otherwise stated, a reference in the sub-contract agreement or in the sub-contract conditions to a clause or schedule is to that clause in, or that schedule to, the appropriate sub-contract agreement or conditions.

Also, unless the context otherwise requires, a reference in a schedule to a paragraph is to that paragraph of that schedule.

3.3.2   Sub-contract to be read as a whole

Clause 1.3 states that the sub-contract is to be read as a whole, but then sets out the priority of documents in the event that there are any inconsistencies.

This is a very important clause and must be used to resolve disputes about which document should be relied upon in the event that there is any inconsistency between the documents.

3.3.3   Headings, references to persons, legislation, etc.

Clause 1.4 clarifies that, unless the context requires otherwise, in the sub-contract agreement, conditions and schedules (as appropriate):

- the headings in the sub-contract are included for convenience only and shall not affect the interpretation of the sub-contract;
- the singular includes the plural and vice versa;
- a gender includes any other gender;

- a reference to a 'person' includes any individual, firm, partnership, company and any other body corporate; and
- a reference to a statute, statutory instrument or other subordinate legislation ('legislation') is to such legislation as amended and in force from time to time, including any legislation which re-enacts or consolidates it, with or without modification.

3.4   Reckoning periods of days

Clause 1.5 makes clear that where an act is required to be done within a specified period of days after or from a specified date, the period shall begin immediately after that date. Where the period would include Christmas Day, Good Friday or a day which under the Banking and Financial Dealings Act 1971 is a bank holiday in England, that day shall be excluded in counting the number of days.

3.5   Contracts (Rights of Third Parties) Act 1999

The Contracts (Rights of Third Parties) Act 1999, which came into force on 11 November 1999, gives a third party the right to enforce a term of a contract *if there is an express provision that he or she should, and if he or she is named in a contract that purports to confer a benefit on him or her.*

Clause 1.6 makes it clear that, notwithstanding any other provision of the sub-contract, nothing in the sub-contract confers or is intended to confer any right to enforce any of its terms on any person who is not a party to it.

Therefore, none of the (un-amended) sub-contracts featured in this book allow for any third-party rights.

3.6   Giving of service of notices and other documents

Clause 1.7.1 says that any notice and other communication expressly referred to in the sub-contract agreement or conditions are to be in writing, and clause 1.7.2 allows the parties to agree upon a format and a means of sending communications.

Clause 1.7.4 says that any notice expressly required by the sub-contract to be given in accordance with clause 1.7.4 shall be delivered by hand or sent by recorded or special delivery post. If sent by recorded or special delivery post, unless it can be proven to the contrary, the notice will be deemed to have been received on the second business day (i.e. any day which is not a Saturday, a Sunday or a public holiday) after the date of posting.

Clause 1.7.3 makes it clear that, subject only to clauses 1.7.2 and 1.7.4 as referred to above, any notice, communication or document may be given or served by any effective means (including by e-mail) and shall be deemed to be duly given or served if addressed and given by actual delivery or sent by pre-paid post to the party to be served at the address stated in the sub-contract agreement or such other address as may from time to time be agreed. If there is no agreed address for the notice to be served, the notice or other document shall be effectively served if given by actual delivery or sent by pre-paid post to the party's last known principal business address, or if a body corporate, its registered or principal office.

Clause 1.7.5 notes that if in an emergency, any communication is made orally with respect to health and safety, risk of damage to property or insurance matters, written confirmation of it shall be sent as soon thereafter as is reasonably practicable.

3.7   Effect of the final payment notice (or the default payment notice)

The effect of the final payment notice is dealt with by way of clause 1.8.

Clause 1.8.1 states that (other than in the case of fraud) the final payment notice shall have effect in any proceedings (whether by adjudication, arbitration or legal proceedings) under or arising out of the sub-contract as conclusive evidence:

■ that where and to the extent that any of the particular qualities of any materials or goods or any particular standard of an item of workmanship was described expressly to be for the approval of the architect/contract administrator, in any of the sub-contract documents, or in any instruction issued by the architect/contract administrator under the main contract that affects the sub-contract works, the particular quality or standard was to the reasonable satisfaction of the architect/contract administrator. However, the final payment notice is not conclusive evidence that the materials or goods or workmanship noted above (or indeed that any other materials or goods or workmanship) comply with any other requirement or term of the sub-contract.

■ that the necessary effect has been given to all of the terms of the sub-contract which require that an amount is to be taken into account in the calculation of the final sub-contract sum. The sub-contracts incorporate what is, in effect, a 'slip rule' in that the final sub-contract sum can be subsequently adjusted where there has been an accidental inclusion or exclusion of any work, materials, goods or figure in any computation or any arithmetical error in any computation. Unfortunately, no guidance is given as to how far this slip rule can extend, but it is submitted that in the context of the significance of clause 1.8, the slip rule would be construed narrowly, and that any accidental inclusion or exclusion would need to be self-evident to an independent third party.

■ that all and only such extensions of time, if any, as are due have been given.

■ that the reimbursement of direct loss and/or expense, if any, due to the sub-contractor is in final settlement of all and any claims which the sub-contractor has or may have arising out of the occurrence of any of the relevant sub-contract matters, whether such claim be for breach of contract, duty of care, statutory duty or otherwise.

■ that the reimbursement of direct loss and/or expense, if any, due to the contractor as a result of the progress of the main contract works being materially affected by any act, omission or default of the sub-contractor is in final settlement of all and any claims which the contractor has or may have so arising, whether such claim be for breach of contract, duty of care, statutory duty or otherwise.

In view of the above, it can be seen that the effect of the final payment notice is far reaching for both parties. It means that if it remains unchallenged, the final payment notice cannot be 'opened up' in respect of the quality of any materials or goods or of the standard of an item of workmanship which had expressly been for the approval of the employer or the architect/contract administrator, the valuation of the sub-contract works, the sub-contractor's entitlement to extensions of time, the sub-contractor's entitlement to direct loss and expense, or the contractor's reimbursement for direct loss and expense.

Therefore, when received, the final payment notice must not be underestimated by sub-contractors.

By way of an interesting *and very important twist,* if the contractor does not issue a final payment notice in accordance with clause 4.12.2, then the sub-contractor may give a default payment notice pursuant to clause 4.12.6, and if the sub-contractor does that, then *all of the provisions set out above (and below) in respect of the final payment notice apply equally to the sub-contractor's default payment notice.*

If the sub-contractor disagrees with the final payment notice (or if the contractor disagrees with the sub-contractor's default payment notice, if applicable), then the dissatisfied party is required to commence adjudication, arbitration or legal proceedings within 10 days after the date of receipt of the final payment notice (or the default payment notice, as applicable). If such proceedings are commenced, the final payment notice or the default payment notice is to have conclusive effect of all matters other than those which are the subject matter of the adjudication, arbitration or legal proceedings. At the conclusion of those proceedings, the effect of the final payment notice (or default payment notice, as applicable) shall be subject to the terms of any decision, award or judgment in or settlement of such proceedings. In respect of adjudication, it was found in one (slightly unusual case[1]) that the adjudication proceedings are commenced by the service of the notice of adjudication rather than by the service of the referral; but, it may be unsafe to rely upon that case as a general precedent, and therefore, it would probably be safer to ensure that the referral is served within the 10-day period stipulated. Notwithstanding the caution noted above, clause 1.8.3 does make

---

[1]*University of Brighton* v. *Dovehouse Interiors Ltd* [2014] EWHC 940 (TCC).

it clear that if an adjudicator gives a decision in respect of proceedings commenced relating to the final payment notice or the default payment notice (as applicable), if either party wishes to have that dispute determined by arbitration or legal proceedings, that party may commence arbitration or legal proceedings within 28 days of the date on which the adjudicator gives his or her decision.

In the situation where:

1. the contractor's employer has commenced proceedings within the periods referred to in clause 1.9 of the main contract, and where such proceedings relate in whole or in part to the sub-contract works, and the sub-contractor is either a party to those proceedings or was made aware of those proceedings within 14 days of them commencing; or

2. the contractor has commenced proceedings against the sub-contractor in respect of the above if the employer's proceedings were commenced after the final payment notice, and provided that the sub-contractor was duly notified of the employer's proceedings, and provided that the contractor commences his or her proceedings against the sub-contractor within 28 days of the employer commencing its own proceedings,

Then the final payment notice or the default payment notice is to have conclusive effect of all matters other than those which are the subject matter of the proceedings noted above. At the conclusion of those proceedings, the effect of the final payment notice (or the default payment notice, as applicable) shall be subject to the terms of any decision, award or judgment in or settlement of such proceedings.

3.8   Applicable law

Clause 1.9 states that the default position is that the sub-contract shall be governed by and construed in accordance with the law of England. If the parties do not wish the law of England to apply, then the parties should make the appropriate amendment to the sub-contract.

3.9   Equivalent sub-contract provisions

In the table below, the equivalent sub-contract provisions to that within the SBCSub/D/A or SBCSub/D/C in the text above are listed for the SBCSub/A or SBCSub/C, DBSub/A or DBSub/C, ICSub/A or ICSub/C and ICSub/D/A or ICSub/D/C.

SBCSub/D/A or SBCSub/D/C	SBCSub/A or SBCSub/C	DBSub/A or DBSub/C	ICSub/A or ICSub/C	ICSub/D/A or ICSub/D/C
Clause 1.1	Clause 1.1	Clause 1.1	Clause 1.1	Clause 1.1
Clause 1.2	Clause 1.2	Clause 1.2	Clause 1.2	Clause 1.2
Clause 1.3	Clause 1.3	Clause 1.3	Clause 1.3	Clause 1.3
Clause 1.4	Clause 1.4	Clause 1.4	Clause 1.4	Clause 1.4
Clause 1.5	Clause 1.5	Clause 1.5	Clause 1.5	Clause 1.5
Clause 1.6	Clause 1.6	Clause 1.6	Clause 1.6	Clause 1.6
Clause 1.7.1	Clause 1.7.1	Clause 1.7.1	Clause 1.7.1	Clause 1.7.1
Clause 1.7.2	Clause 1.7.2	Clause 1.7.2	Clause 1.7.2	Clause 1.7.2
Clause 1.7.3	Clause 1.7.3	Clause 1.7.3	Clause 1.7.3	Clause 1.7.3
Clause 1.7.4	Clause 1.7.4	Clause 1.7.4	Clause 1.7.4	Clause 1.7.4
Clause 1.7.5	Clause 1.7.5	Clause 1.7.5	Clause 1.7.5	Clause 1.7.5
Clause 1.8	Clause 1.8	Clause 1.8	Clause 1.8	Clause 1.8
Clause 1.8.1	Clause 1.8.1	Clause 1.8.1	Clause 1.8.1	Clause 1.8.1
Clause 1.8.3	Clause 1.8.3	Clause 1.8.3	Clause 1.8.3	Clause 1.8.3
Clause 1.9	Clause 1.9	Clause 1.9	Clause 1.9	Clause 1.9
Clause 4.12.2	Clause 4.12.2	Clause 4.12.2	Clause 4.14.2	Clause 4.14.2
Clause 4.12.6	Clause 4.12.6	Clause 4.12.6	Clause 4.14.6	Clause 4.14.6

4 Sub-contractors' General Obligations

4.1 Introduction

In any building sub-contract, the sub-contractor's primary obligation is to carry out and complete the sub-contract works in accordance with the sub-contract documents. This usually takes the form of an express term in the sub-contract to that effect; however, even without such an express term, this obligation would be implied.

The Sub-contractor's obligation to carry out and complete the Works obviously needs to be expanded upon within the Sub-contract. Most sub-contracts set out the general obligations of the sub-contract and the JCT sub-contracts are no different.

Against this background, in this chapter, the following matters will be dealt with, using the sub-contract clause references, etc. contained in the SBCSub-/D/A or SBCSub/D/C (as appropriate). Whilst it is clearly beyond the scope of this book to review every nuance of the other sub-contract forms under consideration, the equivalent provisions (where applicable) within the SBCSub/A or SBCSub/C, the DBSub/A or DBSub/C, the ICSub/A or ICSub/C and the ICSub/D/A or ICSub/D/C are given at the table at the end of this chapter. It must be emphasised that before considering a particular issue, the actual terms of the appropriate edition of the relevant sub-contract should be reviewed by the reader (and/or legal advice should be sought as appropriate) before proceeding with any action/inaction in respect of the sub-contract in question.

In this chapter we will be considering the following:

- General obligations
  - What are the general obligations?
  - What is the sub-contractor's obligation to carry out and complete the sub-contract works?
  - What is meant by the sub-contractor's obligation to carry out and complete the sub-contract works in a proper and workman-like manner?
  - What is the sub-contractor's duty to warn?
  - What are the sub-contract documents that the sub-contractor has to comply with when carrying out and completing the sub-contract works?

The JCT 2011 Building Sub-contracts, First Edition. Peter Barnes and Matthew Davies.
© 2016 John Wiley & Sons, Ltd. Published 2016 by John Wiley & Sons, Ltd.

- What are the sub-contract obligations to carry out and complete the sub-contract works in compliance with the Construction Phase Plan?
- What is the Construction Phase Plan?
- Where are statutory requirements defined in the sub-contract?
- What are the statutory requirements?
- What are the sub-contractor's obligations where he or she sub-lets the works and/or design to another?

■ Sub-contractor's design
- Does the sub-contract include provision for sub-contractor design?
- What are the sub-contractor designed portion works?
- What information is to be provided by the contractor in respect of his or her design obligation?
- In addition to the CDM Regulations, what is stated to be provided by the sub-contractor, and when, under clause 2.6?
- What is the contractor's design submission procedure?
- What is the status of a contractor's design document marked 'A', 'B' or 'C'?
- Must the architect/contract administrator justify why a contractor's design document is marked as status 'B' or 'C'?
- Following his or her receipt what happens if the architect/contract administrator does not respond to a design documents in the required time period?
- What happens subsequently where a drawing is returned marked 'C'?
- Does the architect/contract administrator or the contractor 'approve' the sub-contractor's design documents?
- What if the comments made by the architect/contract administrator upon the sub-contractor's design documents (marked 'C') actually constitute a variation?
- How is the provision of further drawings, details and directions dealt with in relation to sub-contract works and/or the sub-contractor's design?
- Is the sub-contractor responsible for the adequacy of the design contained in the Contractor's Requirements?
- Where is the clause that seeks to protect the sub-contractor in respect of the responsibility of the Contractor's Proposals and the adequacy of the design in this document located within the SBCSub/D/C?
- What is the standard of design expected by the sub-contractor under the sub-contract?
- What is the 'reasonable skill and care' standard of design liability?
- What is the 'fitness for purpose' standard of design liability?
- What is the standard of design liability in the SBCSub/D?
- How does the Defective Premises Act 1972 affect the sub-contractor's design liability?
- Is there a financial limitation of liability in the sub-contract?
- What is 'design development'?
- Errors and failures by the sub-contractor – other consequences

- Materials, goods and workmanship
  - What is the sub-contractor's liability for materials, goods and workmanship?
  - What is workmanship?
  - What is the position in respect of materials and goods?
  - What is the applicable standard of materials and goods?
  - What if the goods are not procurable?
  - What is the standard of workmanship in the SBCSub/D/C?
  - What if the approval of the quality of materials or the standards of workmanship is a matter for the opinion of the architect/contract administrator?
  - What is the Construction Skills Certification Scheme (CSCS)?
- Compliance with main contract and indemnity
  - Is the sub-contractor expressly obliged to comply with any specific obligations the contractor has under the main contract?
  - What is an indemnity clause?
  - Can the sub-contractor's liability be limited?
- Errors, discrepancies and divergences
  - Bills of quantities provided by the contractor
  - What if the bills of quantities are not in accordance with SMM7 or contain errors?
  - How are inadequacies/errors in the Contractor's Requirements dealt with?
  - How is a discrepancy or divergence within the Contractor's Requirements dealt with?
  - What is the position if there is a discrepancy in the Sub-contractor's Design Portion (SCDP) documents (i.e. not the Contractor's Proposals)?
  - Notification of discrepancies
  - Notification of divergence in statutory requirements

4.2   General obligations

4.2.1   What are the general obligations?

Clause 2.1 of SBCSub/D sets out the general obligations of the sub-contractor. These obligations are:

- to carry out and complete the sub-contract works;
- to carry out and complete the sub-contract works in a proper and workman-like manner;
- to carry out and complete the sub-contract works in compliance with the sub-contract documents;
- to carry out and complete the sub-contract works in compliance with the Construction Phase Plan;

- to also carry out and complete the sub-contract works in compliance with other statutory requirements, and to give all notices required by the statutory requirements in relation to the sub-contract works;
- to also carry out and complete the sub-contract works in conformity with directions given in accordance with clause 3.4 plus all other of the Contractor's reasonable requirements (so far as they apply) regulating the carrying out of the main contract works.

4.2.2   What is the sub-contractor's obligation to carry out and complete the sub-contract works?

As noted at the beginning of this chapter, in any building sub-contract, the sub-contractor's primary obligation is to carry out and complete the sub-contract works in accordance with the sub-contract documents.

This usually takes the form of an express term in the sub-contract to that effect (as is the case under SBCSub/D in article 2 and clause 2.1 and the other JCT sub-contracts being considered in this book). Were this obligation not an express term, it would, in any event, be a term implied into the sub-contract.

It is generally accepted, following the judgment in *London Borough of Merton v. Stanley Hugh Leach*,[1] that there is an implied term that a contractor will not hinder or prevent the sub-contractor from carrying out his or her obligations in accordance with the terms of the sub-contract.

In that case, Judge Vinelott said:

'Where in a written contract it appears that both parties have agreed that something should be done which cannot effectively be done unless both concur in doing it, the construction of the contract is that each agrees to do all that is necessary to be done on his part for the carrying out of that thing though there may be no express words to that effect'.

4.2.3   What is meant by the sub-contractor's obligation to carry out and complete the sub-contract works in a proper and workman-like manner?

To the extent that this is not an express clause, it would be implied as a term into the sub-contract.

This means that the sub-contractor must do the work with all proper skill and care.[2] Normally when deciding what level of skill and care is required, the court will consider all the circumstances of the contract, including the degree of skill professed (expressly or impliedly) by the sub-contractor.[3]

Breach of this duty includes the use of materials containing patent defects, even where the sub-contractor had not chosen the source of those materials.

---

[1] *London Borough of Merton v. Stanley Hugh Leach* (1985) 32 BLR 51.
[2] *Young & Marten Ltd v. McManus Childs Ltd* (1969) 9 BLR 77.
[3] *Young & Marten Ltd v. McManus Childs Ltd* (1969) 9 BLR 77.

It may also include relying uncritically on an incorrect plan supplied by the contractor where an ordinarily competent sub-contractor should have had serious doubts about the accuracy of the plan.[4]

This also links into a sub-contractor's duty to warn.

4.2.4   What is the sub-contractor's duty to warn?

Two cases in 1984[5] placed a duty on sub-contractors to warn their employer of design problems that they knew about, and, in certain circumstances, irrespective of whether or not the sub-contractor had any design liability.

In respect of this matter, it has been found by the Court of Appeal that in a case[6] where there was a major roof collapse, the sub-contractor had not done enough to discharge his or her duty of care even though he or she had worked to a design instructed by the client, had discussed the matter with the contractor's engineer and had suggested an alternative solution which was unacceptable to the client. In that case there was a risk of personal injury to the sub-contractor's employees, and this may be why the Court of Appeal considered that the sub-contractor should have protested more vigorously and pressed his or her objections on the grounds of safety, perhaps even to the degree that he or she should have refused to continue to work until the safety of his or her workmen was addressed.

Of course, it is open to question whether the Court of Appeal would have reached the same decision if the sub-contractor's employees were not at risk of personal injury.

With this in mind and in a similar case,[7] where it was alleged that the sub-contractor was under a duty to warn in respect of works to be carried out by others *after he or she had satisfactorily completed his or her work*, the court found that the sub-contractor did *not* have a duty to warn.

The entire question of duty to warn is far from clear, and, in particular, in the situations where there is a design defect that does not amount to something dangerous, and where a sub-contractor should have known of the problem of design, but did not, the law is clearly not yet fully developed.

4.2.5   What are the sub-contract documents that the sub-contractor has to comply with when carrying out and completing the sub-contract works?

SBCSub/D/C defines the sub-contract documents in clause 1.1 as being 'the documents referred to in article 1' of SBCSub/D/A.

---

[4]*Lindenberg* v. *Canning* (1992) 62 BLR 147. In that case, the plan incorrectly showed obviously load-bearing walls as non-load–bearing walls.

[5]*Equitable Debenture Assets Corporation Ltd* v. *William Moss* (1984) 2 Con LR 1; *Victoria University of Manchester* v. *Wilson* (1984–1985) 1 Const LJ 162.

[6]*Plant Construction plc* v. *Clive Adams Associates and Others* [2000] BLR 137, CA.

[7]*Aurum Investments Ltd* v. *Avonforce Ltd and Others* (2001) 17 Const LJ 145.

The documents referred to in article 1 of SBCSub/D/A are:

- the agreement, the sub-contract particulars and the schedule of information;
- the documents referred to in and annexed to the schedule of information;
- the Standard Building Sub-contract conditions SBCSub/D/C (Standard Building Sub-contract Conditions with sub-contractor's design 2011 Edition) incorporating the JCT amendments stated in the sub-contract particulars, and subject to any schedule of modifications included in the numbered documents;
- the numbered documents annexed to the agreement.

4.2.6 What are the sub-contract obligations to carry out and complete the sub-contract works in compliance with the Construction Phase Plan?

As noted above, clause 2.1 requires the sub-contractor to carry out and complete the sub-contract works in compliance with the Construction Phase Plan. The Construction Phase Plan is defined under clause 1.1 as being:

'Those parts of Construction Phase Plan for the Main Contract that are applicable to the Sub-contract Works and annexed to the Schedule of Information, together with any updates and revisions of it by the Principal Contractor notified to the Sub-contractor before or during the progress of the Sub-contract Works'.

4.2.7 What is the Construction Phase Plan?

The Construction Phase Plan is developed in two parts:

- The pre-construction Construction Phase Plan, which is usually put together by the Principal Designer and which brings together the health and safety information from all parties involved at the pre-construction stage. The pre-construction Construction Phase Plan must include information from the employer about inherent risks which reasonable enquiry would reveal, and this plan forms the basis of the development of a construction stage Construction Phase (construction phase plan) by the Principal Contractor.
- The construction-stage Construction Phase, which is developed by the Principal Contractor. This plan is developed to include:
  o   risk and other assessments prepared by the Principal Contractor and other contractors/sub-contractors;
  o   the health and safety policy of the Principal Contractor;
  o   safe method of work statements, etc.

The construction-stage Construction Phase (construction phase plan) forms the basis for the health and safety management of the project and continues to evolve through construction. There is usually a requirement that the construction-stage Construction Phase (previously termed the Health and Safety Plan) is updated at regular intervals.

4.2.8   Where are statutory requirements defined in the sub-contract?

The statutory requirements are defined under clause 1.1 of SBCSub/D/C as being:

'Any statute, statutory instrument, regulation, rule or order made under any statute or directive having the force of law which affects the Sub-contract Works or performance of any obligations under this Sub-contract and any regulation or by-law of any local authority or statutory undertaker which has any jurisdiction with regard to the Sub-contract Works or with whose systems they are, or are to be, connected'.

4.2.9   What are the statutory requirements?

These statutory requirements would include:

- Defective Premises Act 1972
- Health and Safety at Work Act 1974
- Sale of Goods Act 1979
- Supply of Goods and Services Act 1982
- Building Act 1984
- Latent Damage Act 1986
- Insolvency Act 1986
- Consumer Protection Act 1987
- Town and Country Planning Act 1990
- Building Regulations Act 1991
- Housing Grants, Construction and Regeneration Act 1996; as amended by sections 138–145 of part 8 of the Local Democracy, Economic Development and Construction Act 2009
- Party Wall Act 1996
- Scheme for Construction Contracts (England and Wales) Regulations 1998 (SI 1998/649) as further amended, as applicable to the circumstances, by either the (Amendment) (England) Regulations 2011, or (Amendment) (Wales) Regulations 2011 or the (Amendment) (Scotland) Regulations 2011
- Human Rights Act 1998
- Late Payment of Commercial Debts (Interest) Act 1998 (as amended by the Late Payment of Commercial Debts Regulations 2013 ['Regulations']).
- Freedom of Information Act 2000
- Enterprise Act 2002
- Bribery Act 2010

It should be noted that nearly all building work has to comply with the Building Regulations which are updated, amended or changed from time to time, and which stipulate the standards that must be met when carrying out building works.

The Public Health Acts 1875 and 1936 enabled local authorities to make by-laws regulating the construction of buildings. The Public Health Act 1961 provided for the replacement of local building by-laws by the Building

Regulations which, when they came into force in 1966, applied throughout England and Wales with the exception of Inner London, for which the London Building Acts remained in force. The current consolidating statute is the Building Act 1984 and the principal Building Regulations are the Building Regulations 1991 (which came into force on 1 June 1992) and the Building Regulations (Amendment) Regulations 1992 (which came into force on 26 June 1992). The Public Health Act 1936 remains in force in relation to drains and sewers.

Certain buildings are or may be exempt from the Building Regulations, and a requirement of the Building Regulations may be relaxed or dispensed with upon application to the Secretary of State.

The Building Regulations are currently updated on a regular basis, and consist of parts (A)–(P) covering different areas of construction, as follows:

A. Structure
B. Fire safety
C. Site preparation and resistance to contaminants and moisture
D. Toxic substances
E. Resistance to passage of sound
F. Ventilation
G. Hygiene
H. Drainage and waste disposal
I. Combustion appliances and fuel storage systems
J. Protection from falling, collision and impact
K. Conservation of fuel and power
L. Access to and use of buildings
M. Glazing safety in relation to impact, opening and cleaning
N. Electrical safety

Clause 2.1 of SBCSub/D/C makes it an express requirement that the sub-contractor complies with the requirements of the Building Regulations, and therefore a sub-contractor who builds in contravention of the Building Regulations may be in breach of the sub-contract. It is therefore sensible for the sub-contractor to request written confirmation at a pre-contract meeting that all of the required sub-contract works have been approved by building control (building control either being the relevant department of the appropriate local authority, or an equivalent department within an approved external agency [e.g. the NHBC]).

Where a sub-contractor builds in accordance with the contractual design but in contravention of the Building Regulations, it is considered that his or her liability may turn on whether or not he or she was aware of the contravention.[8]

4.2.10 What are the sub-contractor's obligations where he or she sub-lets the works and/or design to another?

The fact that the sub-contractor has sub-let the whole or any part of the sub-contract works (including the design of same) does not affect his or her

---

[8]*Equitable Debenture Assets Corporation Ltd* v. *William Moss* (1984) 2 Con LR 1.

obligations under the sub-contract. Legally, the sub-contractor remains wholly responsible for carrying out and completing the sub-contract works in all respects in accordance with the sub-contract.

In respect of the above, clause 3.2 of SBCSub/D/C makes it clear that the Sub-contractor cannot, without the written consent of the contractor, sub-let the whole or any part of the sub-contract works.

Clause 3.2 of SBCSub/D/C also makes it clear that even where such written consent is given by the contractor, the sub-contractor is to remain wholly responsible for carrying out and completing the sub-contract works in all respects in accordance with the sub-contract, and furthermore, that the contractor's consent to any sub-letting of design must not in any way affect the obligations of the sub-contractor under clause 2.13.1 (which relates to the sub-contractor's design liabilities) of SBCSub/D/C or any other provision of SBCSub/D/C.

4.3   Sub-contractor's design

Design liability can be assumed by either an express term in a sub-contract or, in certain circumstances (despite no express obligation), a sub-contractor may nevertheless assume a design obligation by his or her actions. The latter possibility is a very involved and complicated area of law which is outside the scope of this book.

Some of the JCT sub-contracts considered in this book (i.e. SBCSub/D, DBSub and ICSub/D) can be used in circumstances where a sub-contractor is required to complete the design of all or part of the sub-contract works (when used in conjunction with the appropriate main contract form); others cannot.

Where the JCT sub-contracts do facilitate the option of design by the sub-contractor, the relevant section of the recitals, contract particulars, etc. need completion accordingly.

4.3.1   Does the sub-contract include provision for sub-contractor design?

SBCSub/D does contain provision for sub-contractor's design. The sub-contractor's obligations in respect of the sub-contractor's designed portion (SCDP) are covered under clause 2.2 of SBCSub/D/C, and may be summarized as shown below. The sub-contractor is to:

- Complete the design for the SCDP works (in accordance with the numbered documents to the extent that they are relevant).
- Select any specifications for the kinds and standards of the materials, goods and workmanship to be used in the SCDP works, so far as not stated in the contractor's requirements or the sub-contractor's proposals (clause 2.2.1 of SBCSub/D/C).
- Comply with the directions of the contractor for the integration of the design of the SCDP with the design of the main contract works as a whole,

subject to the provisions of clause 3.5.3 of SBCSub/D/C (clause 2.2.2 of SBCSub/D/C). Clause 3.5.3 of SBCSub/D/C relates to the sub-contractor's objections to an instruction. This matter is covered in Chapter 10.
- Comply with regulations 8–10 of the CDM Regulations (clause 2.2.3 of SBCSub/D/C).

4.3.2   What are the sub-contractor's designed portion works?

The SCDP works are described in the third recital of SBCSub/D/A. This is done either by stating the nature of work in the SCDP works or by referring to the document(s) (i.e. the design document(s)) that more fully describe it.

Further, as the definitions clause (clause 1.1) defines the sub-contractor's designed portion by reference to the third recital (where design responsibility is intended to be transferred to the sub-contractor), the importance of carefully completing the third recital in SBCSub/D/A cannot be overstated.

The process normally followed (in respect of the design documents) is:

- The contractor provides documents (included as numbered documents under sub-contract particulars item 17 of SBCSub/D/A) showing and describing or otherwise stating the requirements of the contractor for the design and construction of the SCDP works. These documents together comprise the contractor's requirements.
- In response to the above, the sub-contractor provides documents (which should also be included as numbered documents under sub-contract particulars item 17 of SBCSub/D/A) showing and describing the proposals of the sub-contractor for the design and construction of the SCDP works. These documents together comprise the sub-contractor's proposals.

4.3.3   What information is to be provided by the contractor in respect of his or her design obligation?

Compliance with CDM Regulations

In the context of the SBCSub/D (and other applicable JCT sub-contracts considered in this book where the Sub-contractor is contracted to undertake design, i.e. DBSub and ICSub/D), where a sub-contractor is responsible for design works, the sub-contractor is a designer.

All of these sub-contracts (as noted above) oblige the sub-contractor to comply with the CDM Regulations (which deal with duties on designers). The designer's duties under the CDM Regulations are discussed further in Chapter 7.

Sub contractor's designed portion works information

Clause 2.6 of SBCSub/D/C details the information that the sub-contractor is required to provide to the contractor in respect of the sub-contractor's designed portion works and also deals with the timing of its supply.

Over and above this, clause 2.6.1 notes that the sub-contractor is also required to comply with regulation 2.2 of the CDM Regulations.

4.3.4   In addition to the CDM Regulations, what is stated to be provided by the sub-contractor, and when, under clause 2.6?

Clause 2.6.1 of SBCSub/D/C obliges the sub-contractor to provide the contractor (free of charge) with three copies of:

- such sub-contractor's design documents and (if requested) calculations as are reasonably necessary to explain or amplify the Sub-contractor's Proposals;
- all levels and setting out dimensions which the sub-contractor prepares or uses for the purposes of carrying out and completing the sub-contractor's designed portion.

Clause 2.6.2 of SBCSub/D/C deals with the timing of the issue of the sub-contractor's design documents.

This sub-clause states that the sub-contractor is to provide the documents 'as and when necessary from time to time' to enable the contractor in respect of the sub-contractor's designed portion to observe and perform his or her obligations:

- under clause 2.9.5 and schedule 1 of the main contract conditions (i.e. the contractor's design submission procedure);
- as otherwise stated in the sub-contract documents (if any other such procedure is stated in the sub-contract documents, these would need to be listed and attached as a numbered document under sub-contract particulars item 17 of SBCSub/A).

Clause 2.6.2 further prohibits the sub-contractor from commencing any work to which a relevant sub-contractor's design documents and other information referred to under clause 2.6.1 until the requirements of the relevant design submission procedure have been satisfied.

The contractor's design submission procedure is discussed next.

4.3.5   What is the contractor's design submission procedure?

In circumstances where a sub-contractor is obliged to provide design documents and other information, the sub-contracts will, invariably, include reference to a design submission procedure.

The contractor's design submission procedure as detailed under schedule 6 of the SBCSub/D/C is as follows:

- The contractor prepares and submits two copies (in such format as is stated in the employer's requirements or the contractor's proposals) of the relevant design documents (including, where appropriate, the sub-contractor's design documents) to the architect/contract administrator in sufficient time to allow any comments of the architect/contract administrator to be incorporated prior to the relevant design document being used for procurement and/or for the carrying out of the works.

No actual time limit is set for the submission of the design documents, but the contractor (and the sub-contractor) need to be aware of the submission procedure to be followed and need to ensure that the various design documents are issued at an early enough date to allow the submission procedure to be completed without causing a delay to the main contract works and/or the sub-contract works.

This is particularly relevant because clause 2.6.2 of SBCSub/D/C makes it clear that the sub-contractor must not commence any work until the relevant design document has satisfactorily completed the submission procedure; clause 2.14 of SBCSub/D/C states that no extension of time will be given to the sub-contractor where he or she has failed to provide in time any necessary design documents in line with clause 2.6.2 of SBCSub/D/C.

■ Within 14 days from the date of receipt of the design documents referred to above (or, if later, 14 days from either the date of or the expiry of the period for submission of the design documents as stated in the contract documents), the architect/contract administrator must return one copy of the contractor's design documents to the contractor marked either 'A', 'B' or 'C'.

4.3.6   What is the status of a contractor's design document marked 'A', 'B' or 'C'?

■ Documents marked with an 'A': If documents are marked with an 'A', that means that the sub-contractor can carry out the work strictly in accordance with those documents.

■ Documents marked with a 'B': If documents are marked with a 'B', that means that the sub-contractor can carry out the work in accordance with the submitted documents, provided that the architect/contract administrator's comments are incorporated and provided that an amended copy of the document in question is promptly submitted to the architect/contract administrator.

■ Documents marked with a 'C': If documents are marked with a 'C', that means that the sub-contractor cannot carry out the work in accordance with the submitted documents without following a further procedure, as outlined next.

4.3.7   Must the architect/contract administrator justify why a contractor's design document is marked as status 'B' or 'C'?

Where design documents are marked with a 'B' or a 'C', the architect/contract administrator is to identify why he or she considers that the document is not in accordance with the main contract (paragraph 4 of the contractor's design submission procedure refers).

4.3.8 Following his or her receipt what happens if the architect/ contract administrator does not respond to a design documents in the required time period?

In such circumstances, following the expiry of the time period, the documents will be regarded as though they had been marked with an 'A'.

Clearly, because of this positive default position, it is vitally important that sub-contractors maintain a strict control over the management and register of documents issued and received.

4.3.9 What happens subsequently where a drawing is returned marked 'C'?

- The sub-contractor is not to carry out any work in accordance with a design document marked with a 'C'.
- The employer (and the contractor, by inference) shall not be liable to make payment for any such work that is executed by the sub-contractor.
- If the sub-contractor agrees with the architect/contract administrator's comments, he or she simply amends the design document and re-submits it to the contractor to allow it to go through the contractor's design submission procedure (as outlined above) again.
- Alternatively, if the sub-contractor does not agree with the architect/ contract administrator's comments, then:
  - the sub-contractor is to notify the architect/contract administrator with reasons (through the contractor), within 7 days of receipt of the architect/contract administrator's comments, why he or she considers that the compliance with the said comments would give rise to a variation.
  - upon receipt of such a notification, the architect/contract administrator shall, within 7 days, either confirm or withdraw the comment.
  - the submission procedure is silent on what happens if a comment is withdrawn. However, by virtue of paragraph 3, it is likely that where the comment is withdrawn by the architect/contract administrator, the design document (previously marked with a 'C') will assume the status of a design document marked with an 'A'.
  - alternatively, where the comment is confirmed by the architect/ contract administrator, the sub-contractor is to amend the design document and is then to re-submit it to the contractor to allow it to go through the contractor's design submission procedure (as outlined above) again.

Of course, if the sub-contractor still does not agree with the architect/ contract administrator's comment, the option of adjudication to resolve the dispute is available to the sub-contractor.

4.3.10   Does the architect/contract administrator or the contractor 'approve' the sub-contractor's design documents?

Nowhere in the above submission procedure does the architect/contract administrator or the contractor 'approve' the sub-contractor's design documents.

The sub-contractor continues to have liability for the sub-contractor's design irrespective of the application of the design submission procedure to it. Any comments by the architect/contract administrator (including, where applicable, the subsequent confirmation or withdrawal by the architect/contract administrator of such comments) under this procedure do not:

■ signify acceptance by the employer or the architect/contract administrator that the design document (or the amended design document, as appropriate) is in accordance with the main contract (or the sub-contract) requirements; or
■ that it gives rise to a variation (refer to SBCSub/D/C, schedule 6, paragraph 8.1).

The sub-contractor's strict liability for design is reinforced by clause 8.3 of schedule 6 (of SBCSub/D/C), which states:

'Neither compliance with the design submission procedure in this Schedule nor with the Architect/Contract Administrator's comments shall diminish the Contractor's obligations to ensure that the Contractor's Design Documents and CDP Works are in accordance with the Main Contract'.

Because of the way that the Contractor's Design Submission Procedure has been incorporated into the SBCSub/D/C, it is submitted that in the above extract, the word 'Contractor' is interchangeable with the word 'Sub-contractor', the CDP Works (contractor's design portion works) must include the sub-contractor's designed works and the Main Contract must incorporate the sub-contract.

On a similar vein, clause 2.13.4 of SBCSub/D/C makes it clear that, although the contractor is to give notice to the sub-contractor specifying anything which appears to him or her to be an inadequacy in the sub-contractor's design documents, no such notice (nor any failure to give such a notice) shall relieve the sub-contractor of his or her obligations in connection with the design.

Finally, and in any event, it is reasonably well established in law that, even if some form of 'approval' were given, this would be unlikely to detract from the sub-contractor's general liability as the designer of the sub-contractor's design works.

4.3.11   What if the comments made by the architect/contract administrator upon the sub-contractor's design documents (marked 'C') actually constitute a variation?

In such circumstances, the sub-contractor needs to be aware that, if the sub-contractor (via the contractor) does not state that he or she considers that the architect/contract administrator's comments constitute a variation (on a design document marked with a 'C') within 7 days of receipt of the architect/contract administrator's comments, then the comments in question shall *not* be treated as giving rise to a variation (refer to SBCSub/D/C, schedule 6, paragraph 8.2). It should be noted that this requirement is a requirement under the main contract; the time period for the sub-contractor may, in effect, need to be less.

4.3.12   How is the provision of further drawings, details and directions dealt with in relation to sub-contract works and/or the sub-contractor's design?

Although, in theory, the numbered documents are intended to show and describe the sub-contract works in full, in practice, there is often the need for further drawings and details to be issued to the sub-contractor, or to be issued by the sub-contractor.

Clauses 2.7.1 and 2.7.2 of SBCSub/D/C deal with this situation. These clauses require:

- the contractor to provide without charge to the sub-contractor:
  - ○ such further drawings or details as are reasonably necessary to explain and amplify the numbered documents; and/or
  - ○ such directions (including those for/or in regard to the expenditure of Provisional Sums) as are necessary to enable the sub-contractor to carry out and complete the sub-contract works in accordance with the sub-contract.
- the sub-contractor to provide to the contractor:
  - ○ working/setting-out drawings and other information necessary for the contractor to make appropriate preparations to enable the sub-contractor to carry out and complete the sub-contract works in accordance with the sub-contract.

Although clause 2.7.2 of SBCSub/D/C does not state that the information to be provided by the sub-contractor is to be free of charge to the contractor, this would probably be the case by implication. Also, as no number of copies is expressly stated, it could be argued that only one copy needs to be provided, although it is more likely that two copies would be implied in keeping with the more general contractor's design submission procedure.

Clauses 2.7.3 and 2.7.4 of SBCSub/D/C require that:

- such further drawings, details, information and directions referred to in clauses 2.7.1 and 2.7.2 shall be provided or given at the time it is reasonably necessary for the recipient party to receive them, having regard to the progress of the sub-contract works and the main contract works;
- where the recipient party has reason to believe that the other party is not aware of the time by which the recipient needs to receive such further drawings, details, information or directions, he or she shall, so far as is reasonably practicable, advise the other party sufficiently in advance to enable him or her to comply with the requirements of clause 2.7 of SBCSub/D/C.

If the sub-contractor has any doubts that the contractor is aware of when the sub-contractor needs to receive such information, the sub-contractor should identify to the contractor what he or she needs, and by when, at his or her earliest opportunity. The same also applies to the contractor, that is, where the date of receipt of the sub-contractor's information noted in clause 2.7.2 affects the execution of work by others.

4.3.13   Is the sub-contractor responsible for the adequacy of the design contained in the Contractor's Requirements?

Until relatively recently, it was commonly thought that a contractor (or sub-contractor) undertaking design did not assume any responsibility for the design (by others) contained within the employer's (or contractor's) requirements. The contractor's design responsibility was limited to the discreet element of the design he or she specifically carried out, that is, the work he or she undertook to finalise the design in the employer's (or contractor's) requirements.

However, in the case of *Co-op* v. *Henry Boot*,[9] it was found that in a standard design and build contract, the contractor was not only responsible for any design specifically carried out by him or her, but more onerously was also liable for the completed scheme as a whole, even where the design work for a specific element was carried out earlier by others.

Accordingly, unless there is a specific clause included in the contract to the contrary, the current position of a sub-contractor undertaking design appears to be that held in *Co-op* v. *Henry Boot*.[10]

However, where a sub-contractor undertakes design, the relevant JCT sub-contracts (i.e. the ones considered in this book) which expressly facilitate some design input from the sub-contractor now include a specific clause which appears largely to protect the sub-contractor from the *Co-op* v. *Henry Boot* position noted above. The SBCSub is no exception.

This clause provides important protection to the sub-contractor.

4.3.14   Where is the clause that seeks to protect the sub-contractor in respect of the responsibility of the Contractor's Proposals and the adequacy of the design in this document located within the SBCSub/ D/C?

Clause 2.8.2 of SBCSub/D/C makes it clear that, subject only to clause 2.12 'divergences from statutory requirements' (dealt with below), the sub-contractor is not responsible for the Contractor's Requirements or for verifying the adequacy of the design contained within them.

Sub-contractors should, however, not assume that such an express term is in their actual sub-contract; rather they must check to see that such an express term is included early in the pre-contract process. This is because the contractor may have amended (or deleted) it within the actual sub-contract he or she is wanting the sub-contractor to sign up to.

---

[9] *Co-operative Insurance Society Ltd* v. *Henry Boot (Scotland) Ltd* [2002] EWHC 1270 (TCC).

[10] *Co-operative Insurance Society Ltd* v. *Henry Boot (Scotland) Ltd* [2002] EWHC 1270 (TCC).

4.3.15   What is the standard of design expected by the sub-contractor under the sub-contract?

If a sub-contractor is required to design some or all of the sub-contract works under the sub-contract, the sub-contractor must identify precisely, pre-contract, what his or her design liability will be under the sub-contract. A crucial point is whether the sub-contractor's design liability is either a 'reasonable skill and care' obligation or a 'fitness for purpose' obligation. The sub-contractor must be aware of this, because legally the differences between these two obligations are fundamental to the sub-contractor's liability.

A 'fitness for purpose' obligation is much stricter than a 'reasonable skill and care' obligation and sets a much higher level of design liability.

In addition to the contract, it needs to be remembered that a designer (including the sub-contractor, where the sub-contractor carries out design work) may also have tortious obligations, but these obligations are outside the scope of this book.

4.3.16   What is the 'reasonable skill and care' standard of design liability?

Most standard form contracts produced by the professional bodies of construction consultants contain an express term which stipulates that the professional will carry out his or her duties using 'reasonable skill and care'.

One of the best definitions for skill and care has been provided by the Court of Appeal in the *Eckersley* v. *Binnie & Partners*[11] case, where Lord Justice Bingham stated:

> 'The law requires of a professional man that he live up in practice to the standards of the ordinary skilled man exercising and professing to have his specialist professional skills. He need not possess the highest expert skill; it is enough if he exercises the ordinary skill of an ordinary competent man exercising his particular art. In deciding whether a professional man has fallen short of the standards observed by ordinary skilled competent members of his profession, it is the standard prevailing at the time of the acts or omissions which prove the relevant yardstick. He is not to be judged by the wisdom of hindsight'.

With regard to the professional man, the standard of reasonable skill and care is also implied by section 13 of the Supply of Goods and Services Act 1982.

4.3.17   What is the 'fitness for purpose' standard of design liability?

Despite the above comments regarding a designer's normal obligation, where there is not an express obligation to the contrary in the sub-contract, the sub-contractor's liability would normally be far more onerous. This liability would

---

[11] *Eckersley* v. *Binnie & Partners* (1988) 18 Con LR 1, CA.

be to meet the standard of 'fitness for purpose', which means that the sub-contractor would be liable to ensure that the finished product was reasonably fit for its intended purpose. Obviously, this is a far higher and more onerous obligation than that normally imposed on professional designers of using 'reasonable skill and care'.

Indeed, where a 'fitness for purpose' design liability applies, it is no defence for the sub-contractor to state he or she has used 'reasonable skill and care' in carrying out his or her design work.

Because of the 'usual' position of fitness for purpose in the absence of an express term, it is not unusual for sub-contracts (and contracts) to include an express term that the liability of the sub-contractor (or main contractor) for design liability is one of 'reasonable skill and care'. Sub-contractors should, however, not assume that such an express term is in their contract; rather they must check to see that such an express term is included early in the pre-contract process.

4.3.18   What is the standard of design liability in the SBCSub/D?

Clause 2.13.1 of SBCSub/D/C makes it clear that when carrying out design work under SBCSub/D/C, the level of care expected of the sub-contractor is one of 'reasonable skill and care' rather than one of 'fitness for purpose'.

As noted above, however, sub-contractors should not assume that such an express term is in their actual sub-contract; rather they must check to see that such an express term is included early in the pre-contract process and has not been amended.

4.3.19   How does the Defective Premises Act 1972 affect the sub-contractor's design liability?

Clause 2.13.2 of SBCSub/D/C provides that where the sub-contract 'involves the sub-contractor in taking on work for or in connection with the provision of a dwelling or dwellings', the design liability of the sub-contractor referred to in clause 2.13.1 of SBCSub/D/C will also include for liability under the Defective Premises Act 1972.

As well as contract and tort, the law imposes a statutory obligation on house builders (and designers thereto). This statute is the Defective Premises Act 1972.

The Defective Premises Act sets down obligations in respect of new homes for occupation. The important aspect of the Act is that anyone designing a new dwelling owes a duty to anyone who later purchases the dwelling to ensure that his or her work is done in a professional manner. Accordingly, under the SBCSub/D, for example, a sub-contractor's liability to such third parties via this statute may be higher than the reasonable skill and care liability he or she owes to the contractor.

However, it should be noted that in order to be in breach of the duty, the failure must cause the dwelling to be unfit for habitation when completed.

4.3.20   Is there a financial limitation of liability in the sub-contract?

Clause 2.13.3 of SBCSub/D/C deals with the financial limitation of liability and states that if a financial limit of liability has been included in the main contract (under clause 2.19.3) in respect of the Contractor's Designed Portion, then that same limit will also apply to the sub-contractor for his or her Contractor's Designed Portion. In the situation where the sub-contractor is not solely liable in respect of this matter, then the sub-contractor would only be expected to make a proportional contribution towards any financial limit set under the main contract.

4.3.21   What is 'design development'?

One of the major areas of dispute in respect of changes to design works relates to what is called 'design development'.

Generally, a sub-contractor may be called upon to design an entire component, such as a cladding system where he or she will be required to consider all relevant matters (e.g. location, wind forces, anticipated snow loadings, etc.), or he or she may be called upon to design to a performance specification (where another designer has already made the basic design considerations and calculations), for example, a central heating system that complies with a performance specification provided.

In the latter case, in particular, if the sub-contractor does not meet the required performance specification, he or she cannot rely on the defence that the product supplied is 'fit for purpose'. In other words, the central heating system designed by the sub-contractor may achieve an acceptable level for the building in question (i.e. it is 'fit for purpose'), but if it does not reach the specification level required by the contractor, then it will be unacceptable.

However, what frequently happens is that a design put forward by a sub-contractor is rejected by a contractor (normally because it has been rejected by the employer) because it is not to the contractor's liking or does not meet the contractor's perceived requirements.

In such a situation, the contractor will often ask for a change to the sub-contractor's design on the basis that this is simply design development (i.e. that the sub-contractor will not get paid any extra for the development to his or her design to meet the requirements of the contractor). The sub-contractor will naturally resist on the basis that the changes required by the contractor are a variation to the scope of the sub-contract works.

In such a situation, it will be for the sub-contractor to demonstrate that his or her submitted design satisfied the requirements of the design brief, or for the contractor to demonstrate that the sub-contractor's submitted design would not satisfy the design brief. Of course, the contractor's objection to the sub-contractor's design would be unsuccessful if it was based upon considerations of price (i.e. savings effected by the sub-contractor that were not being passed onto the contractor) rather than on any technical failings of the sub-contractor's design.

There is not a great deal of case law on this particular subject, but the *Skanska* v. *Egger*[12] Court of Appeal case considered the matter of design development. In that particular case, the parties entered into a contract whereby Skanska agreed, for a so-called guaranteed maximum price, to develop the design of, manage, procure and construct a factory building.

Disputes arose as to various claims by Skanska (in particular in respect of additional steelwork). Skanska argued that the claims arose from changes in the Employer's Requirements, whilst Egger contended that they were merely instances of design development comprehended within the Employer's Requirements.

Skanska was responsible for the completion of the design of the project as outlined in the contract documents and the contract drawings.

The Court of Appeal found that although Skanska was obliged to install more steel than was indicated at tender stage, the tender drawings *did show a requirement for steel* and there was sufficient evidence for the first-instance judge to have decided that the contract provided for more detailed information to be provided post-contract. No further payment was therefore due for the additional work.

The Skanska case illustrates the dangers to a contractor of entering into a design and build contract on a guaranteed maximum price basis where design issues have not been fully formulated at the time of contracting.

Of course, much turns upon the individual contract terms and the facts of a given case, but the principle that a contractor (or a sub-contractor) who takes on design responsibilities may find himself or herself responsible for a good deal of development costs is of universal application.

Finally, sometimes, a sub-contractor is required to carry out a design that is 'to the entire satisfaction' of the contractor (or the employer). This would appear to be an entirely open-ended obligation on the sub-contractor. However, it is submitted that in respect of any such requirement, the contractor's (or the employer's) scope for satisfaction is limited by the design specification provided to the sub-contractor, and that, unless there are any express provisions in the sub-contract to the contrary, the sub-contractor will be entitled to an additional payment, and, if appropriate, additional time if the design level that satisfies the contractor (or the employer) exceeds the design specification level that was initially provided to the sub-contractor. This principle was established in the case of *Dodd* v. *Churton*.[13]

4.3.22 Errors and failures by the sub-contractor – other consequences

Clause 2.14 of SBCSub/D/C states that no extension of time shall be given, and no loss and expense will be allowed, and no (contractual) termination for

---

[12] *Skanska Construction UK Ltd* v. *Egger (Barony) Ltd* [2002] BLR 236.
[13] *Dodd* v. *Churton* [1897] 1 QB 562.

non-payment will be permitted where the cause of the progress of the sub-contract works having been delayed, affected or suspended is:

- any error, divergence, omission or discrepancy in the Sub-contractor's Proposals;
- any error, divergence, omission or discrepancy in any of the SCDP Works information that it is to provide under clause 2.6.1 of SBCSub/D/C;
- failure of the sub-contractor in completing the sub-contractor's design documents to comply with regulation 2.2.3 of the CDM Regulations;
- failure of the sub-contractor to provide in due time any sub-contractor's design documents or related calculations or information either:
  - as required by clause 2.6.2 of SBCSub/D/C (see commentary above); or
  - in response to a written application from the contractor specifying the date (having due regard to the progress of the sub-contract works) when certain relevant documents or other information is required.

4.4   Materials, goods and workmanship

4.4.1   What is the sub-contractor's liability for materials, goods and workmanship?

Before looking at the specific provisions in SBCSub/D/C, there are some common general principles discussed next.

4.4.2   What is workmanship?

The normal understanding of the word 'workmanship' in the construction industry relates to the skill and/or care exercised by a sub-contractor (or a contractor) in the physical execution of the works.

Ordinarily, to the extent that the choice of materials is left to the sub-contractor, 'workmanship' may also mean design (or at least suitability of the materials for the purpose for which they have been used). In the absence of express terms to the contrary, the law has, for many years, implied a term that building work will be carried out in a proper and workman-like manner.[14]

4.4.3   What is the position in respect of materials and goods?

The position generally is that a sub-contractor is to ensure compliance with the specification where materials or goods are specified in the contract. If the specification states a brand name or a particular supplier of the material, the sub-contractor would still be under a warranty that those materials or goods are of good quality when used.[15]

---

[14]*Test Valley Borough Council* v. *Greater London Council* (1979) 13 BLR 63.
[15]Refer generally to the Sale of Goods Act 1979 and the Supply of Goods and Services Act 1982.

A warranty of fitness for purpose (i.e. a warranty that the materials or goods will be fit for the purpose for which they are intended to be used) may also apply if the circumstances indicate that there was a reliance on the sub-contractor's skills regarding the suitability of the materials or goods.[16]

However, where the circumstances indicate that there is no reliance whatsoever on the skill and care on the part of the sub-contractor on the issue of quality (e.g. the terms of the sub-contract compel the sub-contractor to accept particular materials or goods from a particular supplier that, for example, had exclusion clauses in respect of liability), the sub-contractor will not be liable for defects in them (i.e. will not be liable if those materials or goods are not of good quality).[17]

This latter position is, however, very much the exception, and the general position remains that the sub-contractor retains liability for materials and goods that are not of good quality when used, and, where there was a reliance on the sub-contractor's skills regarding the suitability of the materials or goods, the sub-contractor retains liability that the materials or goods are reasonably fit for the purpose for which they will be used.

The reason for this is because the law considers that there is a need to maintain a chain of liability from the employer down to the manufacturer. Without such a chain of liability, a sub-contractor would only be able to recover nominal damages from the next party down the chain (the supplier or manufacturer). Generally, the law adopts the view that society is not well served by allowing those causing loss or damage to escape liability whilst those who suffer the loss are denied any remedy. Wherever possible, therefore, the courts interpret contracts in such a way that the chain of liability is maintained.

Against this background, we can now consider the contractual position under the SBCSub/D/C conditions.

4.4.4   What is the applicable standard of materials and goods?

Clause 2.4.1 of SBCSub/D/C conditions deals with materials and goods and states that:

■ except for the sub-contractor's design works (i.e. SCDP Works), all of the materials and goods for the sub-contract works shall, *so far as procurable*, be in accordance with the kinds and standards described within the sub-contract documents.

■ in respect of the sub-contractor's designed works (i.e. SCDP Works), all of the materials and goods are, *so far as procurable*:
  o   to be of the kinds and standards described in the contractor's requirement; or
  o   where not specifically described in the contractor's requirement, they are to be as described in the Sub-contractor's Proposals or documents referred to in clause 2.6.1.

---

[16] *Young & Marten Ltd* v. *McManus Childs Ltd* (1969) 9 BLR 77.
[17] *Gloucestershire CC* v. *Richardson* (1968) 1 AC 480.

- clause 2.4.1 includes a facility for the sub-contractor to substitute any materials or goods; however, the sub-contractor must receive the written consent of the contractor before doing so. The contractor's consent is not to be unreasonably delayed or withheld. The contractor's consent does not relieve the sub-contractor of his or her other obligations.

Clause 2.4.3 of SBCSub/D/C provides that:

- where and to the extent that approval of the quality or materials or the standards of workmanship is a matter for the opinion of the architect/contract administrator, such quality and standards must be to his or her reasonable satisfaction. This is discussed further in Subsection 4.4.7.
- In addition to the above, clause 2.4.3 of SBCSub/D/C also states that, to the extent that the quality of materials or the standards of workmanship are neither described in the manner referred to in clause 2.4.1 or 2.4.2 nor stated to be a matter for the opinion or satisfaction of the architect/contract administrator, they must be of a standard appropriate to the sub-contract works.

Note also that clause 2.4.4 of SBCSub/D/C requires the sub-contractor (upon the request of the contractor) to provide to the contractor reasonable proof that the materials and goods used by him or her comply with clause 2.4. as appropriate.

The requirement to work to the sub-contract documents implies that, in respect of materials and goods, the fact that the contractor consents to substituted materials or goods does not give the sub-contractor a defence to his or her liability that those materials or goods will be of good quality when used.

From the general principles already explained:

- the sub-contractor is not under a warranty of fitness for purpose in respect of materials or goods specified in the sub-contract documents (but is normally liable that the said materials or goods are of good merchantable quality).

  However, some specifications contain lists of different types of materials or goods for the same purpose, with the choice left to the sub-contractor as to which type to use. In such a case, a sub-contractor would not take on the 'fitness for purpose' liability simply by making a choice of one of the listed materials or goods, since, in such a case, it could not be considered that any reliance had been placed on the sub-contractor's skills regarding the suitability of the materials or goods in question.[18]
- The position is somewhat different to the extent that materials and goods form part of the sub-contractor's designed works (i.e. SCDP Works) as the sub-contractor would almost certainly be liable for both the good quality and the fitness for purpose of those materials and goods.

---

[18]*Rotherham Metropolitan Borough Council v. Frank Haslam Milan & Co Ltd and Others* (1996) 78 BLR 1.

4.4.5   What if the goods are not procurable?

As noted above, clause 2.4.1 of SBCSub/D/C qualifies the sub-contractor's obligation to provide materials or goods as described applies in so far as those materials and goods are procurable.

Because the sub-contract does not specify any geographical limits of procurability, an argument could be raised that materials and goods that are only procurable abroad are still 'procurable' in line with clause 2.4.1 of SBCSub/D/C. Whether or not the materials or goods are procurable (at all) will depend on the facts and what is reasonable.

4.4.6   What is the standard of workmanship in the SBCSub/D/C?

Clause 2.4.2 of SBCSub/D/C states that:

- excluding the sub-contractor's designed works (i.e. the SCDP Works), workmanship for the sub-contract works shall be of the standards described in the sub-contract documents.
- workmanship for the sub-contract's designed works (i.e. the SCDP Works) shall be of the standards described in the Contractor's Requirements or, if not there specifically described, as described in the Sub-contractor's Proposals.

The sub-contractor's obligation on standards of workmanship is not qualified by the phrase 'so far as procurable', which, in theory at least, means that it is no defence to the sub-contractor's obligation that the human skills or equipment to achieve that standard of workmanship are not procurable.

4.4.7   What if the approval of the quality of materials or the standards of workmanship is a matter for the opinion of the architect/contract administrator?

- As noted above, clause 2.4.3 of SBCSub/D/C specifically deals with this issue, and where it applies, states that such quality and standards shall be to the architect's/contract administrator's reasonable satisfaction.
- Obviously, in such a situation, the sub-contractor should obtain confirmation from the contractor of the architect/contract administrator's reasonable satisfaction where it applies to any materials, goods or workmanship in the sub-contract works, as the sub-contract does not provide for any direct communication between the sub-contractor and the architect/contract administrator. It is often appropriate for a sub-contractor to suggest that a sample of his or her work is approved by the architect/contract administrator so that a 'yardstick' for the standard of future works is provided.
- In respect of the architect/contract administrator's reasonable satisfaction, the architect/contract administrator must use an objective standard, and it is therefore submitted that he or she has no power to demand quality or workmanship of the highest standard unless that is reasonable in the circumstances of the case.

■ Another point to note is that the sub-contract refers to materials or goods being to the reasonable satisfaction of the architect/contract administrator and not to the reasonable satisfaction of the contractor.

4.4.8   What is the Construction Skills Certification Scheme (CSCS)?

Clause 2.4.5 of SBCSub/D/C requires the sub-contractor to take all reasonable steps to encourage the sub-contractor's persons to be registered cardholders under the Construction Skills Certification Scheme (CSCS) or to be qualified under an equivalent recognised qualification scheme.

The Construction Skills Certification Scheme (CSCS) is managed by CSCS Ltd. The purpose of the CSCS is to ensure that those involved in the construction industry are competent in their occupation and have health and safety awareness. The CSCS operates by issuing cards (CSCS cards) to those who are qualified to receive the cards, in different colours to suit different occupations.

4.5   Compliance with main contract and indemnity

4.5.1   Is the sub-contractor expressly obliged to comply with any specific obligations the contractor has under the main contract?

Clause 2.5.1 of SBCSub/D/C requires the sub-contractor to observe, perform and comply with the contractor's obligations under the main contract as:

■ identified in or by the schedule of information (included within SBCSub/D/A);
■ in so far as those obligations relate and apply to the sub-contract works (or any part of them).

Those obligations are specifically stated to include (without limitation):

■ those under clause 2.10 of the main contract conditions (levels and setting out);
■ those under clause 2.21 of the main contract conditions (relating to fees or charges legally demandable);
■ those under clauses 2.22 and 2.23 of the main contract conditions (relating to royalties and patent rights);
■ those under clause 3.22 of the main contract conditions (relating to antiquities).

In addition, the sub-contractor is to indemnify and hold harmless the contractor in respect of his or her obligations under the above main contract liabilities, against and from:

■ any breach, non-observance or non-performance by the sub-contractor or his or her employees or agents of any of the provisions of the main contract (refer to clause 2.5.1.1 of SBCSub/D/C);
■ any act or omission of the sub-contractor or his or her employees or agents which involves the contractor in any liability to the employer under the provisions of the main contract (refer to clause 2.5.1.2 of SBCSub/D/C).

Also, under clause 2.5.2 of SBCSub/D/C (subject only to the exceptions contained in insurance clauses 6.4 and 6.7.1 of SBCSub/D/C as dealt with under Chapter 11 of this book), the sub-contractor is to indemnify and hold harmless the contractor against and from any claim, damage, loss or expense due to or resulting from any negligence or breach of duty on the part of the sub-contractor, his or her employees or agents (including any misuse by him/her or them of scaffolding or other property belonging to or provided by the contractor).

As the sub-contract deems the relevant information concerning the main contract by reference to the schedule of information (included within SBCSub-/D/A), it is important that the contractor takes care when completing this information. Likewise, the specific links in this clause to particular clauses in the main contract make it essential that the correct sub-contract published for use with a particular main contract is selected and used.

4.5.2　What is an indemnity clause?

An indemnity clause is a clause where one party agrees to make good a loss suffered by the other party in respect of damage or claims arising out of various matters.

It should be noted that where the SBCSub/D/C contains a general indemnity to the contractor (e.g. clauses 2.5.1 and 2.5.2 of SBCSub/D/C), the sub-contractor's liability could include the contractor's costs, liquidated damages due to the employer and payments to third parties to whom the contractor is liable. Further, being an indemnity, this may increase the liability of the sub-contractor to the contractor for costs, losses and expenses beyond that which might otherwise be recoverable for breach of the sub-contract and is, potentially, unlimited.

4.5.3　Can the sub-contractor's liability be limited?

The contractor and the sub-contractor may expressly agree that the sub-contractor's liability is limited, and this may be done by way of an overall cap on liability of the contractor's costs and damages.

However, there is no provision for such a cap in the sub-contracts considered in this book. If any overall cap on liability is agreed, this would need to be carefully recorded; using SBC/Sub/D as an example, it would need to be recorded on a numbered document listed and attached to sub-contract particulars item 17 of the SBCSub/D/A.

In addition, the indemnity as set out above has the effect of potentially extend-ing the sub-contractor's period of liability because the cause of action (that sets the liability period running) under an indemnity does not generally arise until the loss in question is actually suffered.

4.6　Errors, discrepancies and divergences

Given the volume and complexity of the documentation generally forming part of a building contract, most include express provisions for dealing with any errors or discrepancies between them.

The way that the SBCSub/D deals with these issues is covered in the following sections.

4.6.1   Bills of quantities provided by the contractor

Clause 2.8.1 of SBCSub/D/C indicates that, unless specifically stated otherwise, bills of quantities (including any bills of quantities prepared for the purpose of obtaining a Variation Quotation) must have been prepared in accordance with the Standard Method of Measurement.

Clause 1.1 of SBCSub/D/C defines this as: 'The Standard Method of Measurement of Building Works, 7th Edition … current, unless otherwise stated in the Sub-contract Documents, at the Sub-contract Base Date …'.

4.6.2   What if the bills of quantities are not in accordance with SMM7 or contain errors?

Clause 2.9.1 of SBCSub/D/C refers. If, in respect of any bills of quantities (as referred to in clause 2.8.1), any of the following list applies, the departure, error or omission shall not vitiate (i.e. shall not make invalid or ineffectual) the sub-contract but shall be corrected:

■ The *Standard Method of Measurement of Building Works, 7th Edition* (SMM7) has not been used in the preparation of the bills of quantities, but this departure has not been stated.
■ There is any error in description or quantity.
■ There is any omission of items that should have been measured.
■ There is any error in or omission of information in any item which is the subject of a Provisional Sum for defined work.

Also, where the description of a Provisional Sum for defined work does not provide the information required by SMM7, the description shall be corrected so that it does provide that information.

Clause 2.9.3 of SBCSub/D/C provides that, subject to clause 2.12 of SBCSub-/D/C (notices to be issued by the sub-contractor and directions to be given by the contractor in respect of divergences from statutory requirements), any correction, alteration or modification under clause 2.9.1 of SBCSub/D/C shall be treated as a variation.

4.6.3   How are inadequacies/errors in the Contractor's Requirements dealt with?

Clause 2.9.2 of SBCSub/D/C refers. If an inadequacy is found in any design in the Contractor's Requirements, then if, or to the extent that, the inadequacy is not dealt with in the Sub-contractor's Proposals, the Contractor's Requirements shall be altered or modified accordingly.

Clause 2.9.3 of SBCSub/D/C provides that, subject to clause 2.12 of SBCSub-/D/C, any correction, alteration or modification under clause 2.9.2 of SBCSub-/D/C shall be treated as a variation.

Note that if there is a discrepancy or divergence within the Contractor's Requirements, it is dealt with under clause 2.11.2 of SBCSub/D/C (see later).

4.6.4   How is a discrepancy or divergence within the Contractor's Requirements dealt with?

A discrepancy or divergence is dealt with under clause 2.11.2 of SBCSub/D/C as follows:

- The Sub-contractor's Proposals deal with that discrepancy or divergence:
    - o   In such circumstances, the Sub-contractor's Proposals prevail (on the assumption that they comply with statutory requirements);
    - o   No adjustment will be taken into account in the calculation of the final sub-contract sum; that is, the sub-contract sum (or the sub-contract tender sum) is deemed to include for the resolution of the discrepancy or divergence within the Contractor's Requirements.
- The Sub-contractor's Proposals do not deal with the discrepancy or divergence within the Contractor's Requirements:
    - o   The sub-contractor must inform the contractor in writing of his or her proposed amendment to deal with the discrepancy or divergence within the Contractor's Requirements.
    - o   The contractor will either agree to the proposed amendment or decide how he or she wishes the discrepancy or divergence to be dealt with.
    - o   In either event, the agreement or the decision of the contractor will be treated as a variation.

Of course, the above comments need to be considered against the background of the sixth recital of SBCSub/D/C, which confirms that the contractor has examined the Sub-contractor's Proposals and, subject to the sub-contract conditions, is satisfied that they appear to meet the Contractor's Requirements. It is important to note that the sixth recital does not require the contractor to check that the Sub-contractor's Proposals satisfy the Contractor's Requirements, but merely that 'they appear to meet' them.

4.6.5   What is the position if there is a discrepancy in the Sub-contractor's Design Portion (SCDP) documents (i.e. not the Contractor's Proposals)?

Discrepancies in the sub-contractor's SCDP documents are further dealt with at clause 2.11.1 of SBCSub/D/C.

Clause 1.1 defines the SCDP documents as being 'the Contractor's Requirements, the Sub-contractor's Proposals, the SDP Analysis and the further documents referred to in clause 2.6.1.1'.

Where there are discrepancies or divergences within or between the sub-contractor's SCDP documents (excluding the Contractor's Requirements), the procedure is as follows:

■ The sub-contractor is to give a notice to the contractor (under clause 2.10 as noted below).
■ The sub-contractor is to send a statement setting out his or her proposed amendments to remove the said discrepancy or divergence.
■ The contractor is then to issue his or her directions accordingly.
■ The sub-contractor shall comply with those directions.
■ To the extent that those directions relate to the removal of a discrepancy or divergence, they will not be taken into account in the calculation of the final sub-contract sum.

Clause 2.9.4 of SBCSub/D/C is clear that the final sub-contract sum must not be adjusted to take into account any variations that are issued arising from errors in description or in quantity in the Sub-contractor's Proposals or in the sub-contractor's design works analysis. The same position applies equally for errors relating to the omission of items from these documents.

4.6.6 Notification of discrepancies

Clause 2.10 of SBCSub/D/C deals with 'notification of discrepancies, etc.'
The sub-contractor must immediately give a written notice to the contractor if the sub-contractor finds any departure, error or inadequacy as referred to in clause 2.9 (i.e. Bills of Quantities, etc. as above) or any error or discrepancy in or between any of the following documents:

■ Clause 2.10.1: the sub-contract documents (i.e. the documents referred to in article 1 of SBCSub/A as appropriate);
■ Clause 2.10.2: the main contract;
■ Clause 2.10.3: any direction issued by the contractor under the SBCSub/D/C;
■ Clause 2.10.4: the sub-contractor's designed portion information as detailed under clause 2.6.1 of SBCSub/D/C.

The contractor must then issue a direction as to how the error or discrepancy is to be dealt with.
Whether these constitute a variation or not has been considered in the relevant sections above.

4.6.7 Notification of divergence in statutory requirements

Under clause 2.12 of SBCSub/D/C, if the sub-contractor *or the contractor* becomes aware of any divergence between:

■ the statutory requirements;
■ the documents listed under clause 2.10 of SBCSub/D/C.

then the sub-contractor or the contractor, as appropriate, is required to give a written notice to the other specifying the divergence immediately.

Note however, that in addition to the above, if the divergence is between:

- the statutory requirements; and
- the SCDP documents (defined under clause 1.1 of SBCSub/D/C as being 'the Contractor's Requirements, the Sub-contractor's Proposals, the SDP Analysis and the further documents referred to in clause 2.6.1.1'

then the sub-contractor, at the time of issuing the notice of divergence, is to inform the contractor in writing of his or her proposed amendment for removing the divergence (refer to the '17-day' procedure as follows).

The procedure is then as follows:

1. The '10-day' procedure: The divergence is not between the statutory requirements and the SCDP documents:
   a. Within 10 days of becoming aware, or of receiving a notice from the sub-contractor in respect of a divergence, the contractor must issue directions accordingly.
   b. If the direction(s) varies the sub-contract, the direction(s) is treated as a direction(s) requiring a variation to be issued.
2. The '17-day' procedure: The divergence is between the statutory requirements and the SCDP documents:
   a. Within 17 days of receipt of the sub-contractor's proposed amendment (where applicable) for removing the divergence, the contractor must issue directions accordingly.
   b. The sub-contractor must comply with such directions at no cost.
   c. The only exception is where the divergence is caused by a change to the statutory requirement after the sub-contract base date which necessitates an alteration or modification to the sub-contractor's designed portion. This exception is treated as being a direction requiring a variation to the Contractor's Requirements.

4.7   Equivalent sub-contract provisions

In the table below, the equivalent sub-contract provisions to that within the SBCSub/D/A or SBCSub/D/C in the text above are listed for the SBCSub/A or SBCSub/C, DBSub/A or DBSub/C, ICSub/A or ICSub/C and ICSub/D/A or ICSub/D/C.

SBCSub/D/A or SBCSub/D/C	SBCSub/A or SBCSub/C	DBSub/A or DBSub/C	ICSub/A or ICSub/C	ICSub/D/A or ICSub/D/C
Article 1	Article 1	Article 1	Article 1	Article 1
Article 2	Article 2 Excludes reference to sub-contractor's design	Article 2	Article 2 Excludes reference to sub-contractor's design	Article 2
SCDP works and the Third, Fourth, Fifth and Sixth Recital	Sub-contractor does not design; hence, no equivalent in SBCSub	Sub-contractor's designed works and the Third, Fourth, Fifth and Sixth Recital	Sub-contractor does not design; hence, no equivalent in ICSub	SCDP works and the Third, Fourth, Fifth and Sixth Recital
Sub-contract Particulars item 17		Sub-contract Particulars items and 17		Sub-contract Particulars item 15
Clause 1.1 definitions	Clause 1.1 definitions	Clause 1.1 definitions	Clause 1.1 definitions	Clause 1.1 definitions
Clause 2.1	Clause 2.1	Clause 2.1	Clause 2.1	Clause 2.1 Note clause 2.1 as drafted also includes virtually identical provisions to clause 2.2 in SBCSub/D/C in relation to the Sub-contractor's Designed Portion.
Clause 2.2	Sub-contractor does not design; hence, no equivalent in SBCSub	Clause 2.2	Sub-contractor does not design; hence, no equivalent in ICSub/C	SCDP obligations are covered under clause 2.1 of ICSub/D/C.
Clause 2.4.1	Clause 2.4.1 Excludes any reference to sub-contractor's design as the sub-contractor does not design under this sub-contract	Clause 2.4.1	No specific clause drafted that materials and goods (and workmanship) are to be in accordance with that described in the contract documents. However, clause 2.1 obliges the sub-contractor to carry out and complete the sub-contract works in compliance with the sub-contract documents.	No specific clause drafted that materials and goods (and workmanship) are to be in accordance with that described in the contract documents. However, clause 2.1 obliges the sub-contractor to carry out and complete the sub-contract works in compliance with the sub-contract documents.

(Continued)

SBCSub/D/A or SBCSub/D/C	SBCSub/A or SBCSub/C	DBSub/A or DBSub/C	ICSub/A or ICSub/C	ICSub/D/A or ICSub/D/C
Clause 2.4.2	Clause 2.4.2	Clause 2.4.2	No specific clause equivalent to clause 2.4.2 of SBCSub/D/C, but would be implied. See also clause 2.1, which obliges the sub-contractor to carry out and complete the sub-contract works in compliance with the sub-contract documents.	No specific clause equivalent to clause 2.4.2 of SBCSub/D/C, but would be implied. See also clause 2.1, which obliges the sub-contractor to carry out and complete the sub-contract works in compliance with the sub-contract documents.
N/A	N/A	Clause 2.4.3 Requirements in respect of samples	N/A	N/A
Clause 2.4.3	Clause 2.4.3	Clause 2.4.4 Simply states that where quality of materials or standards of workmanship are not described in the manner stated in clause 2.4.1 or 2.4.2, they must be to a standard appropriate to the sub-contract works. Also see clause 2.4.3 in respect of the obligation to provide samples.	Clause 2.3.1	Clause 2.3.1 Additionally, this clause also deals with the position in respect of the SCDP works too.
Clause 2.4.4	Clause 2.4.4	Clause 2.4.5	N/A	N/A
Clause 2.4.5	Clause 2.4.5	Clause 2.4.6	Clause 2.3.2	Clause 2.3.2
Clause 2.5	Clause 2.5	Clause 2.5	Clause 2.4	Clause 2.4
Clause 2.6.1	Sub-contractor does not design; hence, no equivalent in SBCSub	Clause 2.6.1	Sub-contractor does not design; hence, no equivalent in ICSub/C	Clause 2.5.3
Clause 2.6.2	As above	Clause 2.6.2	As above	Clause 2.6.3

SBCSub/D/A or SBCSub/D/C	SBCSub/A or SBCSub/C	DBSub/A or DBSub/C	ICSub/A or ICSub/C	ICSub/D/A or ICSub/D/C
Schedule 6	Not applicable	Schedule 6	Not applicable	No in-built procedure like SBCSub. Clause 2.6.3 just refers to a design submission procedure in the sub-contract documents. Thus, it relies on the parties agreeing and including one.
Clause 2.7.1	Clause 2.7.1	Clause 2.7.1	Clause 2.5.1	Clause 2.5.1
Clause 2.7.2	Clause 2.7.2	Clause 2.7.2	Clause 2.5.2	Clause 2.5.2 Clause 2.5.3 relates to the issue of the contractor to the sub-contractor's design documents (clause 2.5.3.1) and levels and setting-out dimensions (clause 2.5.3.2).
Clause 2.7.3	Clause 2.7.3	Clause 2.7.3	Clause 2.6.1	Clause 2.6.1
Clause 2.7.4	Clause 2.7.4	Clause 2.7.4	Clause 2.6.2	Clause 2.6.2
Clause 2.8.1	Clause 2.8.1	Clause 2.8.1	Clause 2.7.1	Clause 2.7.1
Clause 2.8.2	Not applicable – sub-contractor does not design	Clause 2.8.2	Not applicable – sub-contractor does not design	Clause 2.21.5
Clause 2.9.1	Clause 2.9.1	Clause 2.9.1	Clause 2.7.2	Clause 2.7.2
Clause 2.9.2	N/A (no design by sub-contractor)	Clause 2.9.2	N/A (no design by sub-contractor)	Clause 2.8.3 Clause 2.8.4
Clause 2.9.3	Clause 2.9.2	Clause 2.9.3	Clause 2.8.2 Clause 2.8.3 Clause 2.9	Clause 2.8.2 Clause 2.8.3 Clause 2.9
Clause 2.10	Clause 2.10 But no reference to SCDP information	Clause 2.10	Clause 2.8	Clause 2.8
Clause 2.11.1	N\A	Clause 2.11.1	N\A	Clause 2.8.3.2 Clause 2.9.1
Clause 2.11.2	N\A	Clause 2.11.2	N\A	Clause 2.8.3 Clause 2.8.4

(Continued)

SBCSub/D/A or SBCSub/D/C	SBCSub/A or SBCSub/C	DBSub/A or DBSub/C	ICSub/A or ICSub/C	ICSub/D/A or ICSub/D/C
Clause 2.12	Clause 2.12 No SCDP works under this Sub-contract; hence, only the 10-day procedure applies	Clause 2.12	Clause 2.10 No SCDP works under this Sub-contract; hence, only the 10-day procedure applies	Clause 2.10
Clause 2.13.1	Not applicable – sub-contractor does not design	Clause 2.13.1	Not applicable – sub-contractor does not design	Clause 2.21.1
Clause 2.13.2	As above	Clause 2.13.2	As above	Clause 2.21.2
Clause 2.13.3	As above	Clause 2.13.3	As above	Clause 2.21.3
Clause 2.13.4	As above	Clause 2.13.4	As above	Clause 2.21.4
Clause 2.14		Clause 2.14		Clause 9 Similar but drafted differently
Clause 3.2	Clause 3.2	Clause 3.2	Clause 3.2	Clause 3.2
Schedule of information (included within SBCSub/D/A)	Schedule of information (included within SBCSub/A)	Schedule of information (included within DBSub/A)	Schedule of information (included within ICSub/A)	Schedule of information (included within ICSub/D/A)

5 Time

5.1 Introduction

Time for performance is a key factor in the JCT sub-contracts (and their corresponding main contracts). The ultimate client (referred to as the Employer under the main contract) wants relative certainty as to when he or she will get his or her building. The commencement date and completion date set may be in the singular (one date) or broken down into various sections depending on the needs of the client; for example, a school may need a section of the building to be completed, allowing it to decant pupils into same, before the next section can be worked on. Each project will have its own particular time needs and constraints, with potential cost implications where a sub-contractor (or a contractor under the main contract) is in delay and has no entitlement under the contract to an extension of time. Sub-contractors need to be aware of the workings of the sub-contract in respect of time, including, but not limited to, the operation of the extension of time provisions therein.

Against this background, in this chapter, the following matters will be dealt with, using the sub-contract clause references, etc. contained in the SBCSub/D/A or SBCSub/D/C (as appropriate). Whilst it is clearly beyond the scope of this book to review every nuance of the other sub-contract forms under consideration, the equivalent provisions (where applicable) within the SBCSub/A or SBCSub/C, the DBSub/A or DBSub/C, the ICSub/A or ICSub/C and the ICSub/D/A or ICSub/D/C are given at the table at the end of this chapter. It must be emphasised that before considering a particular issue, the actual terms of the appropriate edition of the relevant sub-contract should be reviewed by the reader (and/or legal advice should be sought as appropriate) before proceeding with any action/inaction in respect of the sub-contract in question.

In this chapter the following matters are dealt with:

- Time and the adjustment to the period for completion
  - What are the sub-contractor's obligations in respect of time for commencement and completion?
  - Is the sub-contractor entitled to an adjustment of the period for completion?
  - Under what circumstances may a sub-contractor's time period be reduced?
  - Can a relevant sub-contract omission reduce the time period or periods for completion stated in the sub-contract particulars?

The JCT 2011 Building Sub-contracts, First Edition. Peter Barnes and Matthew Davies.
© 2016 John Wiley & Sons, Ltd. Published 2016 by John Wiley & Sons, Ltd.

- ○ What events must occur to allow an adjustment of the time for the sub-contractor's period to be considered?
- ○ What are the relevant sub-contract events?
- ○ What events must occur to allow an adjustment of the time for the sub-contractor's period to be considered?
- ○ Is a sub-contractor required to give notice that it is in delay?
- ○ What is the recipient to do upon receipt of a sub-contractor's notice of delay?
- ○ What happens if a sub-contractor does not issue a notice of delay?
- ○ What factors are to be considered when deciding whether or not there is an entitlement to an extension of time?
- ○ What if there are concurrent delays?
- ○ What measures are a sub-contractor required to take prevent delay?
- ■ Practical completion and lateness
  - ○ Is practical completion defined?
  - ○ What is practical completion?
  - ○ What happens when the sub-contractor is of the opinion that he or she has achieved practical completion?
  - ○ When is practical completion deemed to have been achieved?
  - ○ What happens if the sub-contractor fails to complete on time?

5.2 Time and the adjustment to the period for completion

5.2.1 What are the sub-contractor's obligations in respect of time for commencement and completion?

In line with clause 2.3 of SBCSub/D/C, the sub-contractor is required to commence works in accordance with the programme details stated in the sub-contract particulars.

In respect of the construction works (but not the design works), the contractor is to issue a written notice to commence works to the sub-contractor. Although a written notice to commence the design works is not required by the sub-contract, it would obviously be sensible for such a written notice to be provided in any event.

When it comes to the completion of the works, the sub-contractor is required to carry out and complete the works in accordance with the programme details stated in the sub-contract particulars, and reasonably in accordance with the progress of the main contract works or each relevant section, as appropriate.

The above provisions are subject to the clauses of the sub-contract that deal with the adjustment to the period for completion (these are reviewed further in this Chapter 5 below).

5.2.2 Is the sub-contractor entitled to an adjustment of the period for completion?

Yes, this may be an extension of time or a reduction in time (refer to clauses 2.16–2.19 of SBCSub/D/C).

5.2.3   Under what circumstances may a sub-contractor's time period be reduced?

Under SBCSub/D, a sub-contractor's time period may be reduced:

1. where a pre-agreed adjustment of time (i.e. a reduction in this case) applies:
   - arguably in respect of an agreed variation quotation (its procedure is located at schedule 2); or
   - in respect of an agreed acceleration quotation (its procedure is also located at schedule 2); or
2. because of a relevant sub-contract omission (as defined at clause 2.16.3, i.e. the omission of any work or obligation through a direction for a variation or in respect of a provisional sum for defined work in any bills of quantities has the effect of reducing the sub-contractor's time period).

5.2.4   Can a relevant sub-contract omission reduce the time period or periods for completion stated in the sub-contract particulars?

No. Clause 2.18.6.3 makes it clear that no reduction in time, as noted above under point (2), is to fix a shorter period or periods for completion than that stated in the sub-contract particulars. Therefore, any reduction to the sub-contractor's time period can only be made as a result of a relevant sub-contract omission after an extension of time has previously been granted.

Can a relevant sub-contract omission reduce a previously agreed 'pre-agreed adjustment' *that has extended the sub-contract period or periods for completion stated in the sub-contract particulars?*

In this situation, clause 2.18.6.4 records that no pre-agreed adjustment that extends the sub-contract period shall be subsequently reduced as a result of a relevant sub-contract omission, except for the situation where there is a variation quotation and the relevant sub-contract omission relates *directly* to a the relevant variation that forms the basis of the pre-agreed adjustment.

5.2.5   What events must occur to allow an adjustment of the time for the sub-contractor's period to be considered?

An adjustment of the time for the sub-contractor's period will only be considered in the event that:

- a relevant sub-contract event occurs;
- a pre-agreed adjustment applies (as defined at clause 2.16.2); or
- a relevant sub-contract omission (i.e. the omission of any work or obligation through a direction for a variation or in respect of a provisional sum for defined work) occurs (as defined at clause 2.16.3).

As noted above, a pre-agreed adjustment of time applies in respect of an agreed variation quotation or acceleration quotation (see schedule 2), and a relevant sub-contract omission is an omission of any work or obligation through a direction for a variation or in respect of a provisional sum for defined work.

5.2.6   What are the relevant sub-contract events?

Relevant sub-contract events are listed under clause 2.19 of SBCSub/D/C and these are as follows:

1. Variations – clause 2.19.1.
2. Directions of the contractor given in order to comply with clauses 2.15, 3.15 and 3.16 (excluding an instruction for expenditure of a Provisional Sum for defined work) – clause 2.19.2.1.
3. Directions of the contractor for the opening up or testing of any work, materials or goods under clause 3.17 or 3.18.4 of the main contract conditions, *unless the inspection or test shows that the works, materials or goods were not in accordance with the sub-contract* – clause 2.19.2.2.
4. Directions of the contractor for the opening up or testing of any work, materials or goods under clause 3.10 of the sub-contract conditions, *unless the inspection or test shows that the works, materials or goods were not in accordance with the sub-contract* – clause 2.19.2.3.
5. Deferment of the giving of possession of the site or any section under clause 2.5 of the main contract conditions – clause 2.19.3.
6. Where there are Contract Bills (under the main contract), the execution of work for which an approximate quantity included in those bills is not a reasonably accurate forecast of the quantity of the work required – clause 2.19.4.1.
7. Where there are bills of quantities (supplied by the contractor), the execution of work for which an approximate quantity included in those bills is not a reasonably accurate forecast of the quantity of the work required – clause 2.19.4.2.
8. Antiquities. Compliance with clause 3.22.1 of the main contract conditions and directions related to it. This includes directions passing on instructions under clause 3.22.2 of the main contract conditions (taking steps to preserve any discovered object in the position and location discovered) – clause 2.19.5.
9. The suspension by the sub-contractor under clause 4.11 of any or all his or her obligations under the sub-contract – clause 2.19.6.
10. The suspension by the contractor under clause 4.11 of the main contract conditions of any or all his or her obligations under the main contract – clause 2.19.7.

Clause 3.21 of SBCSub/D/C provides that if the contractor under clause 4.11 of the main contract conditions gives the employer a written notice of his or her intention to suspend the performance of any or all of his or her obligations under the main contract because of non-payment, he or she is to immediately copy that notice to the sub-contractor, and if he or she then suspends his or her obligations under the main contract, he or she is to immediately notify the sub-contractor.

However (see clause 3.22 of SBCSub/D/C), it is most important for the sub-contractor to note that:

    i. the sub-contractor is not to suspend his or her performance in respect of the sub-contract works simply because he or she has been notified by the contractor that the contractor has suspended his or her performance in respect of the main contract works.

    ii. Following the notice from the contractor advising that the contractor has suspended his or her performance in respect of the main contract works, the sub-contractor should (where possible) simply proceed regularly and diligently with his or her sub-contract works until and/or unless he or she receives an express direction from the contractor (under clause 3.22.1 of the sub-contract) to cease the carrying out of the sub-contract works. This cessation is to continue until the contractor directs the sub-contractor to recommence his or her works (in line with clause 3.22.2).

11. Any impediment, prevention or default, whether by act or by omission, by the employer or any of the employer's persons except to the extent caused or contributed to by any default, whether by act or omission, of the sub-contractor or of any of the sub-contractor's persons – clause 2.19.8.

12. Any impediment, prevention or default, whether by act or by omission, by the contractor (including when the contractor is acting as the principal contractor) or any of the contractor's persons except to the extent caused or contributed to by any default, whether by act or omission, of the sub-contractor or of any of the sub-contractor's persons – clause 2.19.9.

13. The carrying out by a statutory undertaker of work in pursuance of his or her statutory obligations in relation to the main contract works, or the failure to carry out such work – clause 2.19.10.

14. Exceptionally adverse weather conditions – clause 2.19.11. It should be noted that this relevant sub-contract event only relates to 'exceptionally adverse' weather conditions. In this context it should be noted that 'exceptional' may be defined as 'unusual' or 'not typical', whilst 'adverse' may be defined as 'un-favourable' or 'harmful'.

The sub-contract does not provide any guidance as to how to judge whether the adverse weather conditions encountered are exceptional. A common approach is to compare the actual adverse weather conditions encountered on site against the records of the previous 5, 10, 20 or 30 years' weather conditions maintained by a meteorological centre that is in the same locality as the site.

15. Loss or damage occasioned by any of the specified perils (as defined under clause 6.1) – clause 2.19.12.

16. Civil commotion or the use or threat of terrorism and/or the activity of the relevant authorities in dealing with such event or threat – clause 2.19.13.

17. Strike, lock-out or local combination of workmen (see clause 2.19.14) affecting any of the:
     i.   trades employed upon the main contract works or any of the trades engaged in the preparation, manufacture or transportation of any of the goods or materials required for the main contract works; or
     ii. persons engaged in preparing the contractor's design for the main contract works.
18. The exercise after the main contract base date by the UK Government of any statutory power which directly affects the execution of the main contract works – clause 2.19.15.
19. *Force majeure* – clause 2.19.16. *Force majeure* is a French law term and this term is generally wider than the common law term act of God. The usual English authority[1] states broadly that it covers all *matters independent of the will of man and which it is not in his power to control.*

5.2.7   Is a sub-contractor required to give notice that it is in delay?

Clause 2.17.1 of SBCSub/D/C requires the sub-contractor to give written notice to the contractor if and whenever it becomes reasonably apparent that the commencement, progress or completion of the sub-contract works is being or is likely to be delayed.

That notice is to provide the material circumstances and the cause or causes of the delay, and is to identify any event which in the sub-contractor's opinion is a relevant sub-contract event.

Clause 2.17.2 of SBCSub/D/C stipulates that when giving this notice, or otherwise in writing as soon as possible afterwards, the sub-contractor is to give particulars of the expected effects of the identified event, and is to give an estimate of the expected delay to the completion of the sub-contract works, and (see clause 2.17.3) is to forthwith notify the contractor in writing should there be any material change in the estimated delay or to any of the particulars provided.

5.2.8   What is the recipient to do upon receipt of a sub-contractor's notice of delay?

When the contractor receives a notice of delay, he or she may request that the sub-contractor provides more information pursuant to clause 2.17.3 of SBCSub/D/C.

Assuming that the contractor is content that he or she has sufficient information from the sub-contractor, then he or she is to take the following action:

■   in line with clause 2.18.1.1 of SBCSub/D/C, he or she is to consider whether the event which is stated to be a cause of delay is a Relevant Sub-contract Event;
■   then (in line with clause 2.18.1.2 of SBCSub/D/C), he or she is to consider whether the said event is likely to delay the completion of the sub-contract

---

[1] *Lebeaupin v. Crispin* [1920] 2 KB 714.

works beyond the period or periods for the completion of the sub-contract work (i.e. those currently fixed either in the sub-contract particulars or as previously revised).

Following this (except where the sub-contract conditions expressly provide otherwise), the contractor is to:

- give an extension of time to the sub-contractor by fixing a revised period or periods for completion as he or she then estimates to be fair and reasonable (clause 2.18.1); or
- he or she is to notify the sub-contractor that he or she does not consider that the sub-contractor is entitled to an extension of time (clause 2.18.2 of SBCSub/D/C).

Whichever response the contractor makes, clause 2.18.2 requires him or her to:

- issue his or her decision as soon as is reasonably practicable;
- issue his or her decision within 16 weeks of receipt of the notice or the required particulars;
- in the case where the date from the receipt of the notice or the required particulars to the date of expiry of the period or periods for the completion of the sub-contract works is less than 16 weeks, then the contractor should endeavour to issue his or her decision prior to the said expiry date.

When issuing his or her decision in respect of an extension of time, in line with clause 2.18.3. of SBCSub/D/C, the contractor is to state:

- the extension of time that he or she has attributed to each Relevant Sub-contract Event; and
- any *reduction in time* that he or she has attributed to a Relevant Sub-contract Omission (i.e. the omission of any work or obligation through a direction for a variation or in respect of a provisional sum for defined work, as fully defined under clause 2.16.3 of SBCSub/D/C).

In respect of the above possible *reduction in time*, the contractor can only take into account any Relevant Sub-contract Omission that occurred since the last occasion when a new period for completion was fixed by the contractor.

Also, although the contractor can reduce the extension of time previously given, clause 2.18.6.3 makes it clear that under no circumstance can the period or periods for completion of the sub-contract works be reduced from that stated in the sub-contract particulars (item 5).

Further, the contractor cannot reduce any extension of time previously given by way of a Pre-agreed Adjustment, unless it applies to a variation quotation where the relevant variation in question related directly to the Pre-agreed Adjustment that was the subject of the Relevant Sub-contract Omission in question (as clause 2.18.6.4).

In addition to the above, clause 2.18.5 notes that if the expiry of the period or periods for completion of the sub-contract works (or of such works in a section) occurs before the sub-contractor's practical completion, the contractor may (not later than the expiry of 16 weeks after the date of the sub-contractor's practical completion):

■ extend the relevant period for completion fixed;
■ shorten the relevant period for completion fixed (if he or she considers that it is fair and reasonable, having regard to any Relevant Sub-contract Omissions issued after the last occasion on which a new period for completion was fixed); or
■ confirm the period or periods for completion previously fixed.

When extending or shortening the relevant period for completion, the contractor is to state:

■ the extension of time that he or she has attributed to each Relevant Sub-contract Event; and
■ any *reduction in time* that he or she has attributed to a Relevant Sub-contract Omission.

5.2.9   What happens if a sub-contractor does not issue a notice of delay?

Clause 2.18.5 of SBCSub/D/C notes that (irrespective of whether or not a notice of delay has been issued by the sub-contractor) if the expiry of the period or periods for completion of the sub-contract works (or of such works in a section) occurs before the sub-contractor's practical completion, the contractor may (not later than the expiry of 16 weeks after the date of the sub-contractor's practical completion):

■ extend the relevant period for completion fixed;
■ shorten the relevant period for completion fixed (if he or she considers that it is fair and reasonable, having regard to any Relevant Sub-contract Omissions issued after the last occasion on which a new period for completion was fixed); or
■ confirm the period or periods for completion previously fixed.

When extending or shortening the relevant period for completion, the contractor is to state (clause 2.18.3):

■ the extension of time that he or she has attributed to each Relevant Sub-contract Event; and
■ any *reduction in time* that he or she has attributed to a Relevant Sub-contract Omission.

5.2.10   What factors are to be considered when deciding whether or not there is an entitlement to an extension of time?

The most important and indeed difficult task in terms of determining delay is to establish the nexus of cause and effect; the mere happening of an event which happens to be a Relevant Sub-contract Event (for example) confers no entitlement to an extension to the contract period.

The test is whether the event in question actually caused a delay to the completion date (or to a section thereof). In considering this matter, it is generally actual progress which must be delayed; the originally planned/programmed progress is not generally relevant.[2]

What needs to be considered are the actual facts on site to determine whether a particular event affected operations on the critical path to the completion date.

What has to be established in determining an extension of time entitlement can briefly be described as follows:

- that an event occurred at all;
- that the event occurred in the manner asserted;
- that the event comes within a category providing an express entitlement to an extension of time within the sub-contract (e.g. that it is a Relevant Sub-contract Event);
- that the required notices have been given; and
- that the event outlined has had a particular delaying effect on the critical path to completion.

5.2.11   What if there are concurrent delays?

In certain instances concurrent delays can occur, that is, two delay events having an impact on the critical path to completion at the same time.

If these two delay events are not the liability of the sub-contractor (or, in fact, if both delay events are the liability of the sub-contractor), then the question of concurrent delays is not a major concern. The difficult issue in respect of concurrent delays occurs when one of the delay events is not the liability of the sub-contractor, but the other delay event is the liability of the sub-contractor. The question then arises as to how the allocation of the delay to completion is allocated between the two concurrent delay events.

Historically, the courts have dealt with this issue in a variety of ways, and some of the more common approaches used are the 'apportionment' approach, the 'American' approach, the 'chronology of events' approach, the 'dominant cause' approach, the application of the 'but for' test and the 'Malmaison'[3] approach (from the English Courts).

Concurrent delay has again, relatively recently, come under the scrutiny of the courts, resulting, it appears in differing views, even between the courts of England and Scotland at the date of writing this book. For background information, this is set out below.

---

[2] *Glenlion* v. *The Guinness Trust* (1987) 39 BLR 89.
[3] Named after the *Henry Boot Construction (UK) Ltd* v. *Malmaison Hotel (Manchester) Ltd case* [1999] 70 Con LR 32.

5.2.11.1   *Current position in Scotland*

In 2010 the Scottish Courts in the City Inn[4] case rejected the Malmasion approach in favour of the apportionment approach, expressed by Lord Osbourne as follows:

> 'where there is true concurrency between a relevant event [entitling the contractor to compensation] and a contractor default, in the sense that both existed simultaneously, regardless of which started first, it may be appropriate to apportion responsibility for the delay between the two causes; obviously, however, the basis for such apportionment must be fair and reasonable.'

5.2.11.2   *Current position in England*

Later, in 2012, this issue then came before the English courts in the Walter Lilley[5] case. On this issue Mr Justice Akenhead rejected the apportionment approach and stated that, in the English jurisdiction, the Malmasion approach applied, restating this as follows:

> '… I am clearly of the view that, where there is an extension of time clause such that as agreed upon in this case and where delay is caused by two or more effective causes, one of which entitles the Contractor to an extension of time as being a Relevant Event, the Contractor is entitled to a full extension of time … although of persuasive weight, the City Inn case is inapplicable within this jurisdiction.'

Clearly therefore, the approach to be adopted in general construction law appears to depend on whether the contract is based upon English Law or Scottish Law. In this regard, and as noted further below, all of the Sub-contracts dealt with in this book are based upon English Law, and therefore, the information above regarding the Scottish Law position is for background information only.

In any event, thankfully, true concurrent delay is rare in practice, as usually, on analysis, the two events are not concurrent and one is the true cause of the delay. Likewise, the position and debate on current delay is still likely to rumble on. The two approaches in two bordering jurisdictions are not ideal. Likewise, no two cases are the same, each being decided on its own unique facts and circumstances applying to that particular project. All of these issues may lead to further cases and further clarification by the courts as this area develops. Therefore, readers need to watch this space and keep abreast of the latest development in this area.

---

[4] *City Inn* v. *Shepherd Construction Ltd* [2010] ScotCS CSIH68.
[5] *Walter Lilly & Company Ltd* v. *Mackay and Another* [2012] EWHC 1773 (TCC) (11 July 2012).

5.2.11.3   Concurrent delay – position under JCT Sub contracts considered in this book

Clause 1.9 of SBCSub/D/C clearly states that the Sub-contract shall be governed by and construed in accordance with the law of England.

The accompanying footnote says that if the parties wish another jurisdiction to apply, appropriate amendments need to be made.

Accordingly, based on case law at the date of writing (which may subsequently change), it appears that under the JCT Sub-contracts being considered (and where the above-mentioned English Law clause has not been amended), the Malmaison position applies. Based on case law at the date of writing, this would appear to suggest that generally if there is a situation where there are two concurrent delay events, one of which a sub-contractor is responsible for (i.e. a culpable delay) and one of which a sub-contractor is not responsible for (i.e. an excusable delay), then the sub-contractor would be entitled to an extension of time (but not necessarily any loss and/or expense).

5.2.12   What measures are a sub-contractor required to take to prevent delay?

Clause 2.18.6.1 of SBCSub/D/C requires the sub-contractor to 'constantly use his best endeavours to prevent delay in the progress of the Sub-contract Works or of such works in any Section, however caused, and to prevent their completion being delayed or further delayed beyond the relevant period for completion'.

Clause 2.18.6.2 of SBCSub/D/C requires the sub-contractor to 'do all that may reasonably be required to the satisfaction of the Contractor to proceed with the Sub-contract Works'.

The wording of clause 2.18.6.1 and clause 2.18.6.2 is fairly typical JCT wording, and there has been much debate over the years as to what the wording 'constantly use his best endeavours to prevent delay' and 'do all that may reasonably be required' means in effect.

Unfortunately, there appears to be no direct case law available to shed any light on this matter. The wording is generally considered to mean that the sub-contractor would be expected to re-programme the works, re-schedule material deliveries and information request schedules, and keep all parties advised of the situation; and it may even mean that the sub-contractor may need to expend some monies to meet this obligation, but it is not normally considered that the sub-contractor would be expected to expend substantial sums of money in respect of this obligation.

Some commentators consider that the wording does not require a sub-contractor to expend any monies at all, but this is probably taking too extreme a view.

5.3   Practical completion and lateness

5.3.1   Is practical completion defined?

The SBCSub/D (and also the SBCSub, the DBSub, the ICSub and the ICSub/D) does not define practical completion.

5.3.2   What is practical completion?

This issue of whether practical completion has been achieved can be a source of dispute between the contracting parties.

Whilst none of the sub-contracts considered in this book (i.e. SBCSub/D, SBCSub, DBSub, ICSub and ICSub/D) define practical completion, the definition of practical completion may be deduced from various court cases, as follows:

1. the works can be practically complete notwithstanding that there are latent defects (i.e. defects that are not apparent at the date of practical completion);
2. a certificate of practical completion may not be issued if there are patent defects (i.e. defects that are apparent at the date of practical completion)[6];
3. practical completion means the completion of all the construction work that has to be done[7];
4. however, a contractor would be expected to have discretion to certify practical completion where there are very minor items of work left incomplete on *de minimis* principles.[8]

5.3.3   What happens when the sub-contractor is of the opinion that he or she has achieved practical completion?

Clause 2.20.1 of SBCSub/D/C requires the sub-contractor to notify the contractor in writing when, in the sub-contractor's opinion, the sub-contract works as a whole or such works in a Section are practically complete and he or she has complied sufficiently with:

■   the requirement to provide as-built drawings, etc.; and
■   the requirement to provide information for the Health and Safety File.

5.3.4   When is practical completion deemed to have been achieved?

Clause 2.20.1 of SBCSub/D/C states that if the contractor does not dissent in writing to the sub-contractor's written notice that practical completion has been achieved, within 14 days of receipt of same, practical completion of such

---

[6]*Jarvis & Sons v. Westminster Corp* [1970] 1 WLR 637; *HW Nevill (Sunblest) v. William Press* (1981) 20 BLR 78.
[7]*Jarvis & Sons v. Westminster Corp* [1970] 1 WLR 637.
[8]*HW Nevill (Sunblest) v. William Press* (1981) 20 BLR 78.

work shall be deemed for all the purposes of this sub-contract to have taken place on the date so notified.

5.3.4.1   What if the contractor does dissent?

The contractor's dissent must be by notice, include reasons for dissenting and be within 14 days of receipt of the sub-contractor's notice – clause 2.20.1.

If the contractor does dissent, then clause 2.20.2 of SBCSub/D/C states that as soon as the contractor is satisfied that the works are practically complete and he or she is satisfied that the sub-contractor has complied sufficiently with the requirement to provide as-built drawings, etc. and the requirement to provide information for the Health and Safety File, then he or she is to notify the sub-contractor in writing of this fact as soon as practicable thereafter, and for all of the purposes of the sub-contract, practical completion will be deemed to have taken place on the date notified by the contractor.

Alternatively, the date that practical completion is achieved may be agreed between the parties, or may be determined by the dispute resolution procedures of the sub-contract.

Clause 2.20.2 of SBCSub/D/C makes it clear that the date of practical completion of the sub-contract works shall not, under any circumstances, be any later than the certified date of practical completion of the main contract works or any relevant Section thereof. It further adds that the contractor is to notify the sub-contractor in writing of any agreed date, which would include the certified date of practical completion of the main contract works or any relevant Section thereof.

5.3.5   What happens if the sub-contractor fails to complete on time?

Clause 2.21 of SBCSub/D/C states that if the sub-contractor fails to complete the sub-contract works within the relevant period for completion, and the contractor gives the requisite notice within a reasonable time after the expiry of that period, the sub-contractor is required to pay the contractor his or her loss and expense resulting from the failure.

This loss and expense may include:

- the contractor's loss and expense;
- the contractor's other sub-contractors' loss and expense; and/or
- where applicable, any liquidated damages suffered by the contractor under the main contract.

An alternative to the sub-contractor paying for the above is for the sub-contractor to allow the amount of the contractor's loss and expense to be deducted from payments otherwise due to him or her.

If the contractor follows this latter option, then it is important that the correct payment and/or payless notices required (as dictated by the circumstances) are put in place by the contractor (in the correct form and issued timeously) as and when stipulated by the sub-contract. Readers are referred to Chapter 8 of this book for more details.

With regard to the above matter, reference should also be made to clause 4.21 of SBCSub/D/C, which deals with the recovery by the contractor of any direct loss and/or expense caused to the contractor where the regular progress of the main contract works is materially affected by an act, omission or default of the sub-contractor, or any of the sub-contractor's persons. Again refer to the Chapter 9 of this book in this book for more details.

5.4   Equivalent sub-contract provisions

In the table below, the equivalent sub-contract provisions to that within the SBCSub/D/A or SBCSub/D/C in the text above are listed for the SBCSub/A or SBCSub/C, DBSub/A or DBSub/C, ICSub/A or ICSub/C and ICSub/D/A or ICSub/D/C.

SBCSub/D/A or SBCSub/D/C	SBCSub/A or SBCSub/C	DBSub/A or DBSub/C	ICSub/A or ICSub/C	ICSub/D/A or ICSub/D/C
Clause 1.9	Clause 1.9	Clause 1.9	Clause 1.9	Clause 1.9
Clause 2.3	Clause 2.3 No reference is made to design work because there is no design obligation upon the sub-contractor under this form of sub-contract.	Clause 2.3	Clause 2.2 No reference is made to design work because there is no design obligation upon the sub-contractor under this form of sub-contract.	Clause 2.2
Clauses 2.16 –2.19 plus schedule 2	Clauses 2.16 –2.19 plus schedule 2	Clauses 2.16 –2.19 plus schedule 2	Clauses 2.12 and –2.13 provide for an extension of time not a reduction to time.	Clauses 2.12 and –2.13 provide for an extension of time not a reduction to time.
Clause 2.19.1 Variations	2.19.1	2.19.1	2.13.1	2.13.1
2.19.2.1 –2.19.2.3 Certain identified contractor's directions	2.19.2	2.19.2	2.13.2	2.13.2
Clause 2.19.3 Deferment of the possession	2.19.3	2.19.3	2.19.3	2.19.3
Clause 2.19.4.1 Contract Bills, and clause 2.19.4.2 Bills of Quantities	Clauses 2.19.4.1 and 2.19.4.2	Clause 2.19.4 Bills of quantities only	Clauses 2.19.4.1 and 2.19.4.2	Clauses 2.19.4.1 and 2.19.4.2

SBCSub/D/A or SBCSub/D/C	SBCSub/A or SBCSub/C	DBSub/A or DBSub/C	ICSub/A or ICSub/C	ICSub/D/A or ICSub/D/C
Clause 2.19.5 Antiquities	Clause 2.19.5	Clause 2.19.5	The equivalent relevant event to SBCSub/D/C 2.19.5 does not appear in ICSub/C	The equivalent relevant event to SBCSub/D/C 2.19.5 does not appear in ICSub/D/C
Clause 2.19.6 suspension by sub-contractor (clause 4.11)	Clause 2.19.6	Clause 2.19.6	Clause 2.13.5	Clause 2.13.5
Clause 2.19.7 suspension by contractor (clause 4.11 of main contract)	Clause 2.19.7	Clause 2.19.7	Clause 2.13.6	Clause 2.13.6
Clause 2.19.8 Impediment, prevention or default (by act or omission) by the employer or any of the employer's persons	Clause 2.19.8	Clause 2.19.8	Clause 2.13.7	Clause 2.13.7
Clause 2.19.9 Impediment, prevention or default, whether by act or by omission, by the contractor or any of the contractor's persons	Clause 2.19.9	Clause 2.19.9	Clause 2.13.8	Clause 2.13.8
Clause 2.19.10 Statutory undertakers	Clause 2.19.10	Clause 2.19.10	Clause 2.13.9	Clause 2.13.9
Clause 2.19.11 Exceptionally adverse weather conditions	Clause 2.19.11	Clause 2.19.11	Clause 2.13.10	Clause 2.13.10
Clause 2.19.12 Specified perils	Clause 2.19.12	Clause 2.19.12	Clause 2.13.11	Clause 2.13.11
Clause 2.19.13 Civil commotion/ Use or threat of terrorism	Clause 2.19.13	Clause 2.19.13	Clause 2.13.12	Clause 2.13.12

(Continued)

SBCSub/D/A or SBCSub/D/C	SBCSub/A or SBCSub/C	DBSub/A or DBSub/C	ICSub/A or ICSub/C	ICSub/D/A or ICSub/D/C
Clause 2.19.14 Strike, lock-out or local combination of workmen affecting main contract trades or design preparation	Clause 2.19.14	Clause 2.19.14	Clause 2.13.13	Clause 2.13.13
Clause 2.19.15 Exercise of statutory power after main contract base date directly affecting the main contract works	Clause 2.19.15	Clause 2.19.15	Clause 2.13.14	Clause 2.13.14
Clause 2.19.16 *Force majeure*	Clause 2.19.16	Clause 2.19.17	Clause 2.13.15	Clause 2.13.15
N/A	N/A	Clause 2.19.16 Late receipt of permission/ approval from the statutory body	N/A	N/A
Clause 2.17.1– 2.17.3 Notice by sub-contractor	Clause 2.17.1 –2.17.3	Clause 2.17.1 –2.17.3	Clauses 2.12.1 and 2.12.4.2	Clauses 2.12.1 and 2.12.4.2
Clause 2.17.3 Provision of further information	Clause 2.17.3	Clause 2.17.3	Clause 2.12.4.2	Clause 2.12.4.2
Clauses 2.18.1 –2.18.6 Fixing period for completion Extension or reduction of time	Clauses 2.18.1 –2.18.6	Clauses 2.18.1 –2.18.6	Clause 2.12 (in particular, clauses 12.2.1 –12.2.3) Similar approach and general intent to clause 2.18 of SBCSub/D/C, but its drafting and application is more simplified Extension of time only – no reduction	Clause 2.12 (in particular, clauses 12.2.1 –12.2.3) Similar approach and general intent to clause 2.18 of SBCSub/D/C, but its drafting and application is more simplified Extension of time only – no reduction

SBCSub/D/A or SBCSub/D/C	SBCSub/A or SBCSub/C	DBSub/A or DBSub/C	ICSub/A or ICSub/C	ICSub/D/A or ICSub/D/C
Clause 2.18.5	Clause 2.18.5	Clause 2.18.5	Clause 2.12.3	Clause 2.12.3
No notice issued by sub-contractor			Up to 16 weeks after the date of practical completion of the sub-contract works or such works in a section, the contractor may make an extension of time whether upon reviewing a previous decision or otherwise and whether or not the sub-contractor has given notice as required by clause 2.12.1.	
Clauses 2.18.6.1 and 2.18.6.2 Sub-contractor's measures required to take prevent delay	Clauses 2.18.6.1 –2.18.6.2	Clauses 2.18.6.1 and 2.18.6.2	Clause 2.12.4.1	Clause 2.12.4.1
Clause 2.20.1	Clause 2.20.1 As the sub-contractor does not design, the requirement to provide as-built drawings found in clause 2.20.1 of SBCSub/D/C is absent.	Clause 2.20.1	Clause 2.14.1 As the sub-contractor does not design, the requirement to provide as-built drawings found in clause 2.20.1 of SBCSub/D/C is absent.	Clause 2.14.1
Clause 2.20.2	Clause 2.20.2	Clause 2.20.2	Clause 2.14.2	Clause 2.14.2
Clause 2.21	Clause 2.21	Clause 2.21	Clause 2.15	Clause 2.15

6 Defects, Design Documents and Warranties

6.1 Introduction

In this chapter the following matters will be dealt with, using the sub-contract clause references, etc. contained in the SBCSub/D/A or SBCSub/D/C (as appropriate). Whilst it is clearly beyond the scope of this book to review every nuance of the other sub-contract forms under consideration, the equivalent provisions (where applicable) within the SBCSub/A or SBCSub/C, the DBSub/A or DBSub/C, the ICSub/A or ICSub/C and the ICSub/D/A or ICSub/D/C are given at the table at the end of this chapter. It must be emphasised that before considering a particular issue, the actual terms of the appropriate edition of the relevant sub-contract should be reviewed by the reader (and/or legal advice should be sought as appropriate) before proceeding with any action/inaction in respect of the sub-contract in question.

In this chapter we will be considering the following areas:

- *Defects*
  - What is a defect?
  - Is a 'defect' defined in the sub-contracts?
  - What is the sub-contractor's liability in respect of defects?
  - Over what period of time is the sub-contractor liable for the rectification of defects?
  - Does the sub-contractor have any liability for defects after the end of the rectification period?
  - What is the latest date that a list of defects can be provided?
  - How often can defects lists be issued and what is the protocol for the clearance of defects?
  - How is the date when the sub-contract defects are made good by the contractor established?
  - Is the sub-contractor's liability for defects limited to only those that are notified to him by the contractor?
  - Can a contractor carry out defects for a sub-contractor and recover the costs incurred?
  - What is an appropriate deduction under the main contract?

The JCT 2011 Building Sub-contracts, First Edition. Peter Barnes and Matthew Davies.
© 2016 John Wiley & Sons, Ltd. Published 2016 by John Wiley & Sons, Ltd.

- *Sub-contractor's design documents*
  - ○ What 'as-built' design documents is a sub-contractor to provide?
  - ○ Does the sub-contract deal with copyright in the sub-contractor's design documents?
- *Collateral warranties*
  - ○ Why do we have collateral warranties?
  - ○ Does the sub-contract oblige the sub-contractor to provide a warranty?
  - ○ What if the sub-contractor wishes to amend the terms of the collateral warranty?
  - ○ Is the collateral warranty to be executed as a deed?
  - ○ Does the sub-contract state that any particular collateral warranty is to be used?
  - ○ What provisions are typically found in a collateral warranty?
  - ○ Does the JCT produce any standard forms of collateral warranty for use with JCT sub-contracts?

6.2 Defects

6.2.1 What is a defect?

A defect is an imperfection, a shortcoming or a failing.

In construction industry terms this may, for example, be paint flaking from walls, a room thermostat not operating correctly or a water pump not operating at all.

6.2.2 Is a 'defect' defined in the sub-contracts?

No. None of the sub-contracts in question define what a defect is.

6.2.3 What is the sub-contractor's liability in respect of defects?

Clause 2.22.1 of SBCSub/D/C requires the sub-contractor to make good at his own cost and in accordance with any direction of the contractor all defects, shrinkages and other faults in the sub-contract works or in any part of them due to materials, goods or workmanship not in accordance with the sub-contract. This liability also extends to any failure of the sub-contractor to comply with his or her obligations in respect of the sub-contractor's designed portion.

6.2.4 Over what period of time is the sub-contractor liable for the rectification of defects?

Clause 2.22 of SBCSub/D/C (see in particular clause 2.22.2) of the sub-contract conditions does not state the period during which the sub-contractor's liability for the rectification of defects applies.

However, it is submitted that this would be the period noted as the rectification period in the main contract (referred to hereafter as the 'rectification period'), which unless specifically stated otherwise, would, by default, be

6 months. It is quite common (particularly where service installations are involved) for the rectification period to be extended to 12 months.

This logic ties in with clause 2.22.2, which amongst other matters, records that the date of the making good of defects by the sub-contractor shall be the date notified by the contractor, with the added long-stop position that this date shall not be later than that date upon which the relevant notice of completion of making good has been issued under the main contract.

6.2.5   Does the sub-contractor have any liability for defects after the end of the rectification period?

Yes. The sub-contractor's liability for defects does not cease upon the expiry of the rectification period; all that ends here is the sub-contractor's right to attend site and make good any defects that are its responsibility under the sub-contract.

In the absence of words to the contrary, the sub-contractor's liability for not completing the sub-contract works (including any applicable design works) in accordance with the sub-contract continues for 6 years from the date of practical completion of the main contract works (in the case of contracts executed under hand) or for 12 years from the date of practical completion of the main contract works (in the case of contracts executed as a deed).

Of course, the sub-contractor's liability for defects which are not related to the sub-contractor not completing the sub-contract works (including the design works) in accordance with the sub-contract ends at the end of the rectification period, and the sub-contractor is not liable for any such defects notified after the end of the rectification period.

However, if a sub-contractor achieves practical completion some considerable time before the contractor achieved practical completion, the sub-contractor's liability for defects could be extended for a period considerably longer than the rectification period (in terms of weeks or months) in respect of the main contract works. This latter point could be particularly significant in the situation where a sub-contractor is relying on manufacturer's warranties to cover for the rectification period of, say, 12 months, when (because of the above situation) the period of liability could considerably exceed 12 months.

6.2.6   What is the latest date that a list of defects can be provided?

Although the sub-contracts in question do not provide any express provision in respect of the latest date that a list of defects can be provided, it is considered that this would be not later than 14 days after the expiry of the rectification period, which is generally consistent with the requirement of the main contract forms.

6.2.7   How often can defects lists be issued and what is the protocol for the clearance of defects?

The SBCSub/D (and also the SBCSub, the DBSub, the ICSub and the ICSub/D) do not stipulate how often defects lists may be provided during the rectification period, nor do they provide for any timetable or protocol for the clearance of the defects, etc. that are listed. It is for this reason that contractors may wish to address (in a numbered document) how defects should be prioritised and the timetable is to be complied with, in respect of such prioritised defects.

6.2.8   How is the date when the sub-contract defects are made good by the contractor established?

Clause 2.22.2 of SBCSub/D/C provides the following steps:

- Sub-contractor notifies the contractor when the sub-contractor is of the opinion that he or she has made good all defects in the sub-contract works (for which he or she has responsibility).
- As soon as the contractor becomes satisfied that the sub-contractor has achieved the making good of all defects in the sub-contract works, the contractor is obliged to notify the sub-contractor accordingly.
- The making good of defects date by the sub-contractor is either:
  - the date notified by the contractor; or
  - a date as agreed or determined by the dispute resolution procedures of the sub-contract; or
  - the making good of defects by the sub-contractor shall be not later than that date upon which the relevant notice of completion of making good has been issued under the main contract.
  - If a date is agreed, this must be notified by the contractor to the sub-contractor.

6.2.9   Is the sub-contractor's liability for defects limited to only those that are notified to him by the contractor?

No. Clause 2.22.3 of SBCSub/D/C makes clear that the sub-contractor's compliance in respect of defects, shrinkages and other faults that are notified to him or her by the contractor shall not in any way limit the sub-contractor's other obligations under clause 2.22.1 (i.e. in respect of all other defects, shrinkages and other faults that are the sub-contractor's responsibility under the sub-contract).

6.2.10   Can a contractor carry out defects for a sub-contractor and recover the costs incurred?

It is generally considered that one of the purposes of defect rectification clauses is to give the sub-contractor the *right* to remedy defects which come within the remit of the clause. Therefore, if a contractor did not give notice to the sub-contractor to clear defects, but simply cleared the defects with his or her own (or other) resources, he or she might not be able to recover any costs at all or he or she might not be able to recover more than the amount that it would have cost the sub-contractor to clear the defects.[1]

---

[1]*Pearce & High Ltd* v. *Baxter* [1999] BLR 101; *William Tomkinson & Sons Ltd* v. *Parochial Church Council of St Michael* (1990) 6 Const LJ 319.

However, the SBCSub/D and the other sub-contracts being considered (i.e. SBCSub, the DBSub, the ICSub and the ICSub/D) do, in certain instances, allow contractors to carry out defects for a sub-contractor in certain instances.

6.2.10.1 SBCSub/D

In respect of the standard building contract main contract conditions then;

- Clause 2.10 allows the Architect/Contract Administrator (with the Employer's consent) to instruct the contractor not to amend any errors arising from the contractor's inaccurate setting out; and
- clause 2.38 allows the Architect/Contract Administrator (with the Employer's consent) to instruct the contractor not to make good defects, shrinkages or other faults that have been notified to the contractor on a defects list.

If such instructions are issued, then both clause 2.10 and clause 2.38 additionally permit an appropriate deduction to be made from the Contract Sum.

Obviously some of the above issues may (or may not) be due to the sub-contractor.

Accordingly, clause 2.23 of SBCSub/D/C notes that where an 'appropriate deduction' has been made under clause 2.10 and/or clause 2.38 of the main contract conditions, then, to the extent that such deduction is attributable to one of the following, the deduction (or an appropriate proportion of the deduction) shall be borne by the sub-contractor and shall be taken into account in the calculation of the final sub-contract sum, or shall be recoverable by the contractor from the sub-contractor as a debt;

- inaccurate setting out by the sub-contractor; or
- defects or other faults in the sub-contract works.

6.2.11 What is an appropriate deduction under the main contract?

Whilst the term 'appropriate deduction' can be commonly found in the JCT suite of contracts (including the Standard Building, Design and Build, and Intermediate main-contracts and sub-contracts), surprisingly, it was not until 2014 that the first case[2] came before the court on the meaning of the term 'appropriate deduction' where a party is instructed not to make good a defect.

The judicial guidance from the court was that each case surrounding the 'appropriate deduction' to be made will depend on its own unique facts and circumstances. At paragraph 30 of the judgment, the court held that it means 'a deduction which is reasonable in all the circumstances and can be calculated by reference to one or more of the following, amongst possibly other, factors:

a. the Contract rates/priced schedule of works/Specification; or
b. the cost to the Contractor of remedying the defect (including the sums to be paid to third party sub-contractors engaged by the Contractor); or

---

[2] *Oksana Mul* v. *Hutton Construction Ltd* [2014] EWHC 1797 (TCC).

c. the reasonable cost to the Employer of engaging another contractor to remedy the defect; or

d. the particular factual circumstances and/or expert evidence relating to each defect and/or the proposed remedial works'.

Clearly the above guidance is wide and open to interpretation based on the individual facts of the specific case in question. What is reasonable in all the circumstances is often a moot point. Factors such as, but not limited to, mitigation of loss by the employer, notification of the defects, access given to the contractor to make good the defect, refusal to return to site, etc. may have a bearing on a particular situation and what is reasonable in the circumstances. Each situation will depend on its own unique facts and circumstances.

In practice, therefore, the potential remains for there to be dispute in the application of an 'appropriate deduction' between the parties, particularly as it is not uncommon for contractors under the main contract to advance the argument that, in the normal course of events, an 'appropriate deduction' would be based upon an abatement to the contractor's value for the works rather than on the cost of completing the defects, etc. by others.

Likewise, the fact that an appropriate proportion of the deduction is to be taken into account in the calculation of the final sub-contract sum may be considered by some to support an argument that the deduction is to be an abatement to value rather than a set-off of costs for completion by others.

6.3 Sub-contractor's design documents

6.3.1 What 'as-built' design documents is a sub-contractor to provide?

Clause 2.24 of SBCSub/D/C requires the sub-contractor to provide to the contractor (at no charge) the following documents before practical completion of the sub-contract works (or such works in any relevant Section) which show or describe the sub-contractor's designed portion as-built information, and which relate to the maintenance and operation of that portion, including any installations forming part of it.

The documents to be provided are:

- such sub-contractor design documents as may be specified in the sub-contract documents; and/or
- such sub-contractor design documents as the contractor may reasonably require.

The above requirement does not detract from the sub-contractor's obligations under clause 3.20.4 of SBCSub/D/C in respect of the Health and Safety File.

Clearly it is in the interest of the sub-contractor to identifying what design documents are necessary and be in a position to issue this information *prior* to practical completion.

6.3.2   Does the sub-contract deal with copyright in the sub-contractor's design documents?

Clause 2.25 of SBCSub/D/C deals with the rights/licences conferred upon the contractor and the employer in respect of the copyright in the sub-contractor's design documents.

It should be particularly noted that the listed rights are conditional upon the sub-contractor being paid all monies due under the sub-contract in full (see clause 2.25.2).

The copyright in all of the sub-contractor's design documents remain vested in the sub-contractor.

However, the contractor shall have an irrevocable, royalty-free, non-exclusive licence, with the full right to sub-licence the employer, to copy, use and reproduce the sub-contractor's design documents for any purpose relating to the main contract works, including, without limitation, those listed as follows:

- the construction of the main contract works;
- the completion of the main contract works;
- the maintenance of the main contract works;
- the letting of the main contract works;
- the sale of the main contract works;
- the promotion of the main contract works;
- the advertisement of the main contract works;
- the reinstatement of the main contract works;
- the refurbishment of the main contract works; and
- the repair of the main contract works.

The contractor also has an irrevocable, royalty-free, non-exclusive licence, with the full right to sub-licence the employer, to copy *but not to reproduce* (which must means using the sub-contractor's design documents as the basis of a similar design issue) the sub-contractor's design documents for any *extension* of the main contract works.

Clause 2.25.3 makes it clear that the sub-contractor is not liable if the employer and/or the contractor uses the sub-contractor's design documents for any purpose other than that they were prepared for.

6.4   Collateral warranties

6.4.1   Why do we have collateral warranties?

It is normally far easier to pursue an action in contract than it is to pursue an action in tort, and it is for this reason that the employer will normally require a collateral warranty to be in place.

A collateral warranty is a separate agreement that establishes a direct contractual link between a sub-contractor (particularly one that has a design liability) and the employer (and/or other interested parties).

6.4.2   Does the sub-contract oblige the sub-contractor to provide a warranty?

Clause 2.26 of SBCSub/D/C notes that where part 2 of the main contract particulars (as annexed to the schedule of information) and/or other sub-contract document provides for the giving by the sub-contractor of collateral warranties to the following, then the sub-contractor is to execute and deliver a collateral warranty within 14 days from receipt of a notice from the contractor that identifies the beneficiary and requires execution of a collateral warranty:

- a purchaser;
- a tenant or funder; and/or
- the employer.

6.4.3   What if the sub-contractor wishes to amend the terms of the collateral warranty?

Clause 2.26.1 of SBCSub/D/C states that if the sub-contractor requires any (reasonable) amendment to the proposed collateral warranty, he or she is to notify the contractor within 7 days of receipt of the proposed collateral warranty, of the proposed amendments. The contractor would then obviously need to seek approval for the proposed changes from the employer.

Pursuant to clause 2.26.1 of SBCSub/D, the sub-contractor is thereafter to execute and deliver the collateral warranty (with any approved amendments, as applicable) within 7 days of being notified of the decision of the employer in respect of the request made for any amendments (irrespective of whether the requested amendments were agreed to or not).

6.4.4   Is the collateral warranty to be executed as a deed?

Clause 2.26.2 of SBCSub/D/C makes it clear that where the main contract is executed as a deed (i.e. where a 12-year limitation of liability period applies), the collateral warranties are to be executed as a deed.

Likewise, where the main contract is executed under hand (i.e. where a 6-year limitation of liability period applies), the collateral warranties 'may' be executed under hand (the word 'may' here denotes that these may be executed either under hand or as a deed).

6.4.5   Does the sub-contract state that any particular collateral warranty is to be used?

The SBCSub/D (and the SBCSub, DBSub, ICSub and ICSub/D) does not stipulate any particular form of collateral warranty to use.

However, the SBCSub guide notes that, if the collateral warranties are not the standard collateral warranties that the JCT produces, then copies of the required forms of collateral warranties should be annexed to the schedule of information forming part of the sub-contract (i.e. the SBCSub/D/A, SBCSub/A, DBSub/A, ICSub/A and ICSub/D/A).

Obviously, if the sub-contractor has any queries regarding, or requires any changes to the terms of the proposed collateral warranty (whether this be the standard warranty or an ad hoc warranty), these should be raised by the sub-contractor before the sub-contract is entered into, since, if they are not, it will be more difficult for the sub-contractor to insist on the required changes at a later date.

6.4.6   What provisions are typically found in a collateral warranty?

Collateral warranties normally consist of the following parts:

1. The parties:
    (a) The warrantor—i.e. the sub-contractor in this case.
    (b) The warrantor's employer—i.e. the contractor in this case.
    (c) The beneficiary—i.e. the purchaser and/or a tenant, and/or the funder, and/or the employer.
2. Recitals
    These set the scene so as to enable anyone reading the document to understand its background. They should briefly describe the project, the roles of the parties and the reason for the warranty being entered into.
3. The warranty
    This is the core and most important part of the warranty. It usually begins with words such as 'The [warrantor] warrants as at and with effect from . . . [e]' and then goes on, by means of a number of sub-clauses, to set out the different obligations of the warrantor.
    The most common warranty is that the warrantor has and will continue to perform his obligations under whichever contract the warranty is in respect of.
    The normal effect of such warranties is to repeat for the beneficiary (e.g. the employer) the obligations that the warrantor (i.e. the sub-contractor) has to his or her client (i.e. the contractor).
4. Supplementary warranties
    Matters covered by supplementary conditions, which generally should not be required, include:
    (a) That the warrantor has exercised reasonable skill and care in the selection of materials (a list of deleterious materials is usually included).
    (b) That the works designed by the warrantor will satisfy any performance specification contained in the contract with the warrantor's client.
    (c) That the works will be of sound manufacture and workmanship, and (for design and build contracts) will satisfy the requirements of the employer.
5. Qualification
    Collateral warranties are often qualified by a number of provisions. It is common to see a provision that the warrantor shall have no greater

liability to the beneficiary, nor a liability of longer duration, than the warrantor would have had if the beneficiary had been named as the warrantor's client in the principal contract to which the warranty refers. The purpose of such a qualification is to ensure that the warranty does not extend the underlying obligations. This is important to the warrantor as any extension of his or her liabilities may fall outside the limits of his or her insurances.

6. Professional indemnity insurance

    Collateral warranties usually contain provisions that require the warrantor to maintain professional indemnity insurance. The level of cover will be specified and it will also be specified that it shall be maintained for 6 or 12 years from practical completion (normally of the main contract works), provided that the insurance is available at commercially reasonable rates and on reasonable terms. This proviso is sufficient protection to the warrantor, and it is not necessary to qualify the requirement by asking the warrantor to use best or reasonable endeavours to maintain the insurance. The warrantor will often be required to produce evidence that the insurance is being maintained and renewed.

    One issue that can arise in relation to professional indemnity insurance is whether the insurance covers each and every claim or if it is on an aggregate basis for all claims within any year. Most beneficiaries will be advised not to accept aggregate insurance, as there is a danger that there will be no cover for some claims made in the same year as others because of an aggregate limit in value being reached.

    It should also be borne in mind that many professional indemnity insurance policies are provided on a 'claims made' basis, which means that the applicable insurance is that which covers the period when the claim was made, not that which covers the period when the alleged defect occurred.

    In respect of professional indemnity insurance, it is in the interest of both the warrantor and the beneficiary that the terms of the warranty fall within the terms of the warrantor's policy. In this regard it should be noted that the purpose of the professional indemnity insurance is to provide a fund of monies upon which the beneficiary can call in the event that a claim needs to be made. If the terms of the warranty do not fall within the terms of the warrantor's policy, that fund of money will not exist.

    A reality of the construction industry is that many smaller contractors and professional consultants have very limited assets and, in the absence of satisfactory insurance, would be unable to meet any significant claim under a warranty.

    It is, therefore, in the interests of both the warrantor (i.e. the subcontractor) and the beneficiary that the terms of the warranty fall within the risks usually covered by a professional indemnity insurance policy.

    The warrantor obviously does not wish to be faced with claims that he or she cannot meet (or which, if imposed, would put him or her out of business), and the beneficiary seeks the comfort of knowing that a source of accessible funds is available. Common sense dictates that it is necessary to strike a balance between these two demands.

7. Step-in rights
    Such provisions are popular where the beneficiary is a funder or freehold purchaser. The purpose of 'step-in rights' is to allow the beneficiary to take over the contract with the warrantor in the event that either the warrantor has a right to terminate that contract or if the beneficiary's own contract has been breached.

8. Copyright licence
    Collateral warranties, as a general rule, contain an irrevocable royalty-free copyright provision in favour of the beneficiary. Such a licence enables the beneficiary to reproduce and use the warrantor's design information. The rights of the beneficiary are normally limited to the use of the information for purposes related to the development.

9. Contracts (Rights of Third Parties) Act 1999
    The Contracts (Rights of Third Parties) Act is almost invariably excluded in most standard and bespoke collateral warranty forms.

There are two other important points to be aware of:

1. An important factor with collateral warranties, from the warrantor's viewpoint, is that he or she should have no greater liability to the beneficiary than he or she has to his or her client and he or she has the same defences against a claim from the beneficiary as he or she would have against his or her client.

2. It is also important to note that the terms of a collateral warranty that a party to a contract is required to give to a third party must be known and/or available at the time the principal contract is formed. In the event that this is not the case, then there is unlikely to be sufficient certainty over the particular terms requiring the giving of a collateral warranty, making such a requirement difficult to enforce.

6.4.7   Does the JCT produce any standard forms of collateral warranty for use with JCT sub-contracts?

In respect of the SBCSub/D, SBCSub, DBSub, ICSub and ICSub/D, the normal situation would be that the standard collateral warranties that the JCT produces would be used. In which case, many of the above issues (where applicable) would be automatically covered.

The three standard collateral warranties produced by the JCT are:

- the Sub-contractor Collateral Warranty for a Funder (SCWa/F);
- the Sub-contractor Collateral Warranty for a Purchaser or Tenant (SCWa/P&T); and
- the Sub-contractor Collateral Warranty for an Employer (SCWa/E).

The said collateral warranties are each provided with a three-page Guidance Notes section.

It should be noted that, in respect of the JCT standard collateral warranties, the section of the collateral warranty that relates to professional indemnity insurance should be deleted, and the warranty particulars "professional indemnity insurance" entry of "does not apply" selected and used, if the sub-contractor has no design liability.

6.5 Equivalent sub-contract provisions

In the table below, the equivalent sub-contract provisions to that within the SBCSub/D/A or SBCSub/D/C in the text above are listed for the SBCSub/A or SBCSub/C, DBSub/A or DBSub/C, ICSub/A or ICSub/C and ICSub/D/A or ICSub/D/C.

SBCSub/D/A or SBCSub/D/C	SBCSub/A or SBCSub/C	DBSub/A or DBSub/C	ICSub/A or ICSub/C	ICSub/D/A or ICSub/D/C
Clause 2.22.1	Clause 2.22.1 No reference is made to defects in any sub-contractor's designed portion as this does not apply under this sub-contract	Clause 2.22.1	Clause 2.16.1 No reference is made to defects in any sub-contractor's designed portion as this does not apply under this sub-contract	Clause 2.16.1
Clause 2.22.2	Clause 2.22.2	Clause 2.22.2	Clause 2.16.2	Clause 2.16.2
Clause 2.22.3	Clause 2.22.3	Clause 2.22.3	Clause 2.16.3	Clause 2.16.3
Clause 2.23	Clause 2.23	Clause 2.23	Clause 2.17	Clause 2.17
Clause 2.24	N/A No sub-contractor design under this sub-contract.	Clause 2.24 Identical obligation in respect of Sub-contractor's Designed Works.	N/A No sub-contractor design under this sub-contract.	Clause 2.19
Clause 2.25.1	N/A No sub-contractor design under this sub-contract	Clause 2.25.1	N/A No sub-contractor design under this sub-contract	Clause 2.20.1
Clause 2.25.2	N/A	Clause 2.25.2	N/A	Clause 2.20.2
Clause 2.25.3	N/A	Clause 2.25.3	N/A	Clause 2.20.3
Clause 2.26.1	Clause 2.26.1	Clause 2.26.1	Clause 2.18.1	Clause 2.18.1
Clause 2.26.2	Clause 2.26.2	Clause 2.26.2	Clause 2.18.2	Clause 2.18.2
Part 2 of main contract particulars (annexed to the Schedule of Information)	Part 2 of main contract particulars (annexed to the Schedule of Information)	Part 2 of main contract particulars (annexed to the Schedule of Information)	Part 2 of main contract particulars (annexed to the Schedule of Information)	Part 2 of main contract particulars (annexed to the Schedule of Information)

7 Control of the Sub-contract Works

7.1 Introduction

As with the performance of any contract, there are certain issues that must be considered and effective controls need to be in place to deal with same. Health and Safety is an obvious important area, but there are other issues too for which the sub-contract needs to have effective provisions in place that deal with such matters.

Against this background, in this chapter, the following matters will be dealt with, using the sub-contract clause references, etc. contained in the SBCSub/D/A or SBCSub/D/C (as appropriate). Whilst it is clearly beyond the scope of this book to review every nuance of the other sub-contract forms under consideration, the equivalent provisions (where applicable) within the SBCSub/A or SBCSub/C, the DBSub/A or DBSub/C, the ICSub/A or ICSub/C and the ICSub/D/A or ICSub/D/C are given at the table at the end of this chapter. It must be emphasised that before considering a particular issue, the actual terms of the appropriate edition of the relevant sub-contract should be reviewed by the reader (and/or legal advice should be sought as appropriate) before proceeding with any action/inaction in respect of the sub-contract in question.

The matters dealt with in this chapter are:

- *Assignment and sub-letting*
  - What is assignment?
  - Can the sub-contractor assign the sub-contract?
  - Does the sub-contractor need to get consent to sub-let any works?
  - Are there any conditions for sub-letting?
- *Person-in-charge*
  - What is the obligation regarding a person-in-charge?
- *Access provided by the sub-contractor*
  - Must the sub-contractor provide access for the contractor, employer and architect/contract administrator?
- *Opening up the works and remedial measures*
  - Can the contractor issue instructions/directions to open up and test?
  - Is the sub-contractor paid for complying with directions to open up and test?
  - What if work is found not to be in accordance with the sub-contract ('non-compliant work')?

The JCT 2011 Building Sub-contracts, First Edition. Peter Barnes and Matthew Davies.
© 2016 John Wiley & Sons, Ltd. Published 2016 by John Wiley & Sons, Ltd.

- ○ What about instructions or deductions under the main contract?
- ○ What about workmanship issues?
- ○ What about non-compliant work by others?
- ○ Indemnity by the sub-contractor
- ○ Schedule 5 – the Sub-contract Code of Practice
- ■ *Attendance and site conduct*
  - ○ What attendances are to be provided?
  - ○ By the sub-contractor?
  - ○ By the contractor?
  - ○ What about other attendance items not listed in SBCSub/D/A?
  - ○ Who can use the scaffolding on site?
  - ○ What does the sub-contract say about possible misuse of scaffolding, etc.?
- ■ *Temporary buildings*
  - ○ Who determines the location of sub-contractor's site accommodation?
  - ○ What are reasonable facilities?
  - ○ Joint Fire Code
- ■ *Site clearance*
  - ○ How is the sub-contractor's waste dealt with?
- ■ *Health and safety and CDM*
  - ○ Health and Safety
  - ○ What is the relevant health and safety legislation?
- ■ *The CDM Regulations (i.e. the Construction [Design and Management] Regulations 2015)*
  - ○ Does the sub-contractor get paid for complying with CDM 2015?
  - ○ What is the purpose of the CDM 2015 Regulations?
  - ○ What are the sub-contractors' duties under the CDM Regulations?
  - ○ When do the CDM Regulations apply?
  - ○ When must a project be notified to the Health and Safety Executive (HSE)?
  - ○ What are the employer's duties under CDM 2015?
  - ○ What are the designer's duties under CDM 2015?
  - ○ What are the Principal Designer's duties under CDM 2015?
  - ○ What are the Principal Contractor's duties under CDM 2015?
  - ○ What are the contractor's duties (and sub-contractor's) under CDM 2015?
  - ○ How does the role of the Principal Contractor affect the sub-contractor?
  - ○ How does health and safety competence affect sub-letting?
  - ○ What is pre-construction information?
  - ○ What is the Construction Phase Plan?
  - ○ What is the Health and Safety File?
- ■ *Suspension of the main contract by the contractor*
  - ○ the General position
  - ○ Does the sub-contract provide for the sub-contractor to be notified of the contractor's proposed suspension/suspension of his or her performance under the main contract?
  - ○ What should the sub-contractor do when notified that the contractor has suspended his or her performance under the main contract?

○   In what circumstances should the sub-contractor cease the sub-contract works when the main contract is suspended by the contractor?
■ *Other provisions – strikes*
    ○   What happens when the main contract works are affected by strikes, etc.?
■ *Benefits under the main contract*
■ *Certificates/statements or notices under the main contract*
    ○   How does the sub-contractor receive this information?

7.2   Assignment and sub-letting

7.2.1   What is assignment?

A contract contains both benefits and burdens for both parties; one party's benefit is usually the other party's burden.

In law, the general principle is that benefits under a contract may be assigned (i.e. legally transferred from the original contracting party to a third party) without the consent of the other party, but burdens under a contract may not. Therefore, without the consent of the contractor, a sub-contractor could not normally assign his or her liability to complete, as this is a burden, not a benefit.[1]

Strictly speaking, if both the benefit and the burden of a contract are legally transferred to a third party, then that is a novation, and not an assignment. A novation is a tripartite agreement by which an existing contract between party X and party Y is legally discharged and a new contract, usually on the same terms as the first contract, is made between party X and (third) party Z.

7.2.2   Can the sub-contractor assign the sub-contract?

Clause 3.1 of SBCSub/D/C states that the sub-contractor shall not assign the sub-contract or any rights thereunder without the written consent of the contractor. Clause 1.7.1 of SBCSub/D/C records the requirement for such consent to be in writing.

Against the above background, the reference in clause 3.1 of SBCSub/D/C that the sub-contractor shall not assign the sub-contract obviously does not, in truth, make a great deal of sense.

However, in the two consolidated appeals of *Linden Gardens* v. *Lenesta Sludge and St Martins* v. *Sir Robert McAlpine*,[2] Lord Browne-Wilkinson said:

> 'Although it is true that the phrase "assign this contract" is not strictly accurate, lawyers frequently use those words inaccurately to describe an assignment of the benefit of a contract since every lawyer knows that the burden of a contract cannot be assigned'.

---

[1]For this principle, see *Nokes* v. *Doncaster Amalgamated Collieries Ltd* [1940] AC 1014.
[2]*Linden Gardens* v. *Lenesta Sludge Disposals* [1993] 3 WLR 408; *St Martins Property Corporation Ltd* v. *Sir Robert McAlpine Ltd* [1993] 3 All ER 417.

In the light of the above, what clause 3.1 of SBCSub/D/C does is to make clear by express terms that, irrespective of the general principle of law, no benefit (or right) (in addition to any burden) under the sub-contract may be assigned to any third party without the written consent of the contractor.

7.2.3   Does the sub-contractor need to get consent to sub-let any works?

Subject to an express term contained in a contract, the general law usually permits a sub-contractor to sub-let his or her contractual obligations to a third party for their performance (termed vicarious performance) on the strict basis that the sub-contractor still remains fully liable and responsible to the contractor under the sub-contract for all his or her contractual obligations.

When the above position applies, it does not matter who performs the work, provided that they have the relevant skill and expertise to do so.

However, if the obligation to be undertaken is of a personal nature (i.e. the particular skill and expertise of an individual sub-contractor is being relied upon to perform the contractual obligations) then the general law will not permit sub-letting.

Clause 3.2 of SBCSub/D/C states that the sub-contractor shall not, without the written consent of the contractor, sub-let the whole or any part of the sub-contract works (clause 3.2.1), or sub-let the design for the sub-contractor's designed portion (clause 3.2.2).

In neither instance is the contractor's consent to be unreasonably delayed or withheld.

It should be noted that these clauses do not say that the contractor's consent is *not* to be delayed or withheld, but that such consent is not to be *unreasonably* delayed or withheld.

What is unreasonable will depend on the particular individual circumstances that exist in relation to the proposed sub-subcontractor; that is, it will be considered on a case-by-case basis. There are no hard and fast rules. Typically, some influencing circumstances may include, but not be limited to: the sub-subcontractor's relevant competence and technical expertise, size of the organisation and available resources to do the work efficiently and on time, their financial standing, track record, safety record, general reputation, etc.

By way of example, if a contractor had serious doubts about the ability of a proposed sub-subcontractor (i.e. a proposed sub-contractor to the sub-contractor) to deal with a particular aspect of the sub-contract works, or of the sub-contractor's designed portion, then it may be entirely reasonable for him or her to delay his or her consent to the use of the proposed sub-subcontractor until he or she is provided with documentary evidence (perhaps references, etc.) that he or she may have requested the sub-contractor to provide to him or her, before he or she gives his or her consent to the use of the proposed sub-subcontractor.

Alternatively, if the contractor had received or was receiving a very poor service from the proposed sub-subcontractor on another project, and therefore had serious doubts that the proposed sub-subcontractor would be capable of carrying out the sub-contract works (or a part thereof) to the required standard, then it may be entirely reasonable for him or her to withhold his or her consent to the use of the proposed sub-subcontractor entirely.

Note also that clause 3.2 of SBCSub/D/C makes it quite clear that, even if the contractor does give his or her written consent to the sub-contractor to sub-let the whole or any part of the sub-contract works, the sub-contractor is to remain wholly responsible for carrying out and completing the sub-contract works in all respects in accordance with the sub-contract.

In addition to the foregoing, note that the contractor's consent to any sub-letting of design shall not in any way affect the obligations of the sub-contractor under clause 2.13.1 of SBCSub/D/C or any other provision of SBCSub/D.

7.2.4   Are there any conditions for sub-letting?

Clause 3.3 of SBCSub/D/C provides only two specific conditions of sub-letting. These are:

- Clause 3.3.1: the sub-subcontractor's employment under the sub-subcontract shall terminate immediately upon the termination (for any reason) of the sub-contractor's employment under the sub-contract.
- Clause 3.3.2 is in two parts as follows:
  - In relation of the main contract works and the site, both the sub-contractor and the sub-sub-contractor undertake to one another to comply with the CDM Regulations.
  - The sub-contractor shall provide for simple interest to be paid to the sub-subcontractor at the level of and subject to the terms equivalent to those of clauses 4.10.6 and 4.12.7 of SBCSub/D/C in respect of any payment, or any part of it, not properly paid to the sub-subcontractor by the final date for payment. The interest rate stated in the SBCSub/D being a rate 5% per annum above the official dealing rate of the Bank of England current at the date that a payment due under the sub-contract becomes overdue.

Apart from the two qualifications that must be included in any sub-letting arrangement, a sub-contractor is at liberty to agree any other terms (that are enforceable) with his or her sub-subcontractor as he or she wishes.

Whilst the JCT has published a sub-subcontract form for use in such a situation (denoted as SubSub), the sub-contractor is not obliged to use that particular form when sub-letting.

7.3   Person-in-charge

7.3.1   What is the obligation regarding a person-in-charge?

Clause 3.8 of SBCSub/D/C requires the sub-contractor to ensure that at all times during the execution of the sub-contract works, he or she has on site a competent person-in-charge. Clause 3.8 also adds that any directions given by the contractor to that competent person-in-charge will be deemed to have been issued to the sub-contractor.

7.4   Access provided by the sub-contractor

7.4.1   Must the sub-contractor provide access for the contractor, employer and architect/contract administrator?

Clause 3.9 of SBCSub/D/C obliges the sub-contractor to allow access at all reasonable times to the contractor, the employer and to any person authorised by either the contractor or the employer to any work which is being prepared for or is to be utilised in the sub-contract works.

The only restrictions to this requirement being those that are necessary to protect any proprietary rights. This latter situation may arise where a product involves particular trade secrets and/or the manufacturer wishes to prevent the manufacturing process from becoming known to third parties.

7.5   Opening up the works and remedial measures

The procedures to be followed in respect of the opening up of works and the remedial measures to works that may be required after the works have been opened up are set out as follows.

7.5.1   Can the contractor issue instructions/directions to open up and test?

Clause 3.10 of SBCSub/D/C allows the contractor to issue directions requiring the sub-contractor to:

- open up for inspection any work covered up;
- arrange for or carry out any test of any materials and goods (whether or not already incorporated in the sub-contract works); or
- arrange for or carry out any test of any executed work.

7.5.2   Is the sub-contractor paid for complying with directions to open up and test?

Clause 3.10 of SBCSub/D/C states that the sub-contractor is to be paid the cost of any such opening up or testing directed (including the cost of any resultant making good) except where the opening up or testing:

- is included in the sub-contract sum; or
- shows that the materials, goods or works were not in accordance with the sub-contract requirements.

In the above circumstances, such cost are borne by the sub-contractor.

If the work is not defective, the sub-contractor may also be entitled to an extension of time (clause 2.19.2.3) and loss and/or expense (clause 4.20.2.3).

7.5.3 What if work is found not to be in accordance with the sub-contract ('non-compliant work')

If any work, materials or goods are found not to be in accordance with the sub-contract (termed 'non-compliant work'), the contractor, in addition to his or her other powers, may:

- under clause 3.11.1 of SBCSub/D/C, issue directions requiring the removal from site or rectification of all or any of the non-compliant work, with or without similar instructions having been issued under clause 3.18 of the main contract conditions. Note that if a similar instruction under clause 3.18 of the main contract conditions:
  - ○ Has not been issued: the contractor must, before issuing the direction, consult with the sub-contractor and give due regard to the Sub-contract Code of Practice set out in schedule 5 to the SBCSub/D/C.
  - ○ Has been issued: the contractor's direction issued under this clause must state this.
- under clause 3.11.2 of SBCSub/D/C, after consultation with the sub-contractor, issue directions requiring a variation as is reasonably necessary in consequence of a clause 3.11.1 direction. Clause 3.11.2 states that provided such directions are reasonably necessary, the sub-contractor will *not* be entitled to payment, nor an extension of time, as a consequence of the direction.
- having due regard to the Sub-contract Code of Practice set out in schedule 5 to the SBCSub/D/C, issue such directions under clause 3.10 of SBCSub/D/C to open up for inspection or to test (as is reasonable in all of the circumstances) to establish to the contractor's reasonable satisfaction the likelihood or extent (as appropriate to the circumstances) of any further similar non-compliance (refer to clause 3.11.3 of SBCSub/D/C). Provided such directions are reasonable, the sub-contractor will not be paid for the opening up, testing or subsequent making good irrespective of whether the opening up and/or testing shows the work to be compliant or non-compliant. If the works are found to be compliant, the sub-contractor may, however, be entitled to an extension of time (i.e. if works directed delay the completion date of the sub-contract works or any Section) as per clause 2.19.2.3. Note that the sub-contractor may also be able to recover any resultant (proven) loss and expense (refer to clause 4.20.2.3 of SBCSub/D/C) if the works are found to be compliant.

It should be noted that the Sub-contract Code of Practice is enclosed in schedule 5 of SBCSub/D/C.

Despite the above powers of the contractor, the contractor may not issue any of the above directions in respect of materials or goods or workmanship where the approval of the quality and standards is a matter for the opinion of the architect/contract administrator, unless such direction comprises or reflects an instruction of the architect/contract administrator under the main contract which relates to the sub-contract works (refer to clause 2.4.3 of SBCSub/D/C).

Where the non-compliant work is the subject of an Architect/Contract Administrator's instruction under clause 3.18 of the main contract conditions, the contractor is to notify the contractor of this as part of issuing his or her directions to the sub-contractor.

7.5.4 What about instructions or deductions under the main contract?

7.5.4.1 Clause 3.18.2 of the SBC main contract

If any work, materials or goods are not in accordance with the contract, then the Architect/Contract Administrator may consult with the contractor and, with the agreement of the employer, may allow all or any of such work, materials or goods to remain (except for those that form part of the contractor's designed portion); and an appropriate deduction would then be made to the ascertained final (main contract) sum.

7.5.4.2 Clause 3.12 of SBCSub/C

If under clause 3.18.2 of the main contract conditions:

- the consultation relates (in whole or in part) to the sub-contract works, the contractor must immediately consult with the sub-contractor (clause 3.12.1 of SBCSub/D/C). This appears to imply that this should be before the employer makes his or her decision.
- When the Architect/Contract Administrator allows all or any non-compliant work to remain, the contractor must notify the sub-contractor in writing and (where applicable) an appropriate deduction would then be made to the final sub-contract sum or, if this was not possible, then the amount of such deduction is recoverable by the contractor from the sub-contractor as a debt (clause 3.12.2 of SBCSub/D/C).

Clause 3.12.3 of SBCSub/D/C requires the sub-contractor to comply with any contractor's directions requiring a variation as are reasonably necessary in consequence of the employer allowing all or any non-compliant work to remain. Like clause 3.11.2 of SBCSub/D/C (discussed at section 7.5.3 of this Chapter), provided such directions are reasonably necessary, the sub-contractor is neither paid for the variation nor entitled to an extension of time.

7.5.5 What about workmanship issues?

Clause 3.13 of SBCSub/D/C deals with the situation where 'workmanship' rather than simply 'work' is not in accordance with the sub-contract.

Accordingly, where there has been a failure to carry out work:

- in a proper and workman-like manner; and/or
- in accordance with the Construction Phase Plan;

the contractor, in addition to his or her other powers, may, after consultation with the sub-contractor, issue such directions (whether requiring a variation or otherwise) as are in consequence reasonably necessary. Where the failure to comply is the subject of an instruction issued by the employer under clause 3.19 of the main contract conditions, the contractor must inform the sub-contractor of this fact when issuing his or her directions (clause 3.13.2 of SBCSub/D/C).

Provided such directions are reasonably necessary, the sub-contractor is not entitled to payment or an extension of time or loss and expense in respect of the variation.

7.5.6 What about non-compliant work by others?

Under clause 3.14.1 of SBCSub/D/C, the sub-contractor must comply with the contractor's directions to take down, re-execute, re-fix and/or re-supply any sub-contract work that has been properly executed, or materials or goods properly fixed or supplied under the sub-contract where necessitated due the removal or rectification of non-compliant work by others (i.e. by the contractor or *another* sub-contractor). Such directions, however, must be issued before the date of practical completion of the sub-contract works or of such works in any relevant Section.

Where such directions are issued, then:

■ in line with clause 3.14.1 of SBCSub/D/C, a copy of the direction must be sent to the sub-contractor whose non-compliant work or failure to comply gave rise to the direction being issued; and
■ under clause 3.14.2 of SBCSub/D/C, the sub-contractor is entitled to:
  ○ payment for such work directed on the basis of a fair valuation;
  ○ an extension of time (assuming that the works in question caused a delay to the completion date of the sub-contract works or such works in any Section);
  ○ to recover any consequential (and proven) loss and expense.

7.5.7 Indemnity by the sub-contractor

In the situation where there is non-compliant work by the sub-contractor or there is a failure by him or her to carry out work in accordance with the sub-contract, clause 3.15 of SBCSub/D/C requires the sub-contractor to indemnify the contractor in respect of any liability and any costs the contractor incurs as a direct result of such non-compliance and/or such failure by the sub-contractor to carry out work in accordance with the sub-contract.

7.5.8 Schedule 5 – the Sub-contract Code of Practice

SBCSub/D/C refers to the Sub-contract Code of Practice and this is referred to when directions in respect of clauses 3.11.1 and 3.11.3 of SBCSub/D/C are being considered by the contractor. Although the contractor is not bound to follow the code of practice, if a dispute arose, his or her case would be greatly weakened unless he or she could show that the terms of the code had been fully considered before his or her directions were issued. The purpose of the code is to ensure that directions under clauses 3.11.1 and 3.11.3 of SBCSub/D/C are not issued unreasonably.

A review of the sub-contract code of practice is outside the scope of this book.

7.6   Attendance and site conduct

7.6.1   What attendances are to be provided?

7.6.1.1   By the sub contractor?

Clause 3.16.1 of SBCSub/D/C states that all items of attendance, other than those listed either:

■ in item 6 of the sub-contract particulars (SBCSub/D/A); and/or
■ in a numbered document referred to in sub-contract particulars (item 17 of SBCSub/D/A)

are required to be provided by the sub-contractor.

7.6.1.2   By the contractor?

The sub-contract particulars in item 6.1 of SBCSub/D/A provides a basic list of attendance items that will be provided by the contractor free of charge. This list is as follows:

■ Provision and erection of all necessary scaffolding or other access/work equipment for all work at height except where its provision is 'necessary solely and exclusively for the purpose of carrying out the Sub-contract works.' This seeks to give due consideration to the Work at Height Regulations 2005. At the time of writing the Health and Safety Executive (HSE) guide entitled *Working at Height: A Brief Guide* states that 'Work at height means work in any place where, if there were no precautions in place, a person could fall a distance liable to cause personal injury' and that *'If you are an employer or you control work at height* (for example if you are a contractor or a factory owner), the Regulations apply to you'. Whilst outside the scope of this book (this regulation should be referred to direct), obligations are placed on Employers and those who have control of any work at height activity to assess the risks and to ensure that such work is correctly planned, supervised and undertaken by competent people, using the correct equipment for working at height.

General legal duties are also applicable to Employees to take reasonable care of themselves and all others who may be affected by their actions, and furthermore to co-operate with their employer to enable their health and safety duties and requirements to be complied with.
■ Space for temporary site accommodation: This does not, however, cover the accommodation itself – see clause 3.17 of SBCSub/D/C.

- Single-phase supply of electricity at 240 V for temporary site accommodation: The actual connection from that point of supply to the temporary site accommodation is to be carried out by (or at the expense of) the sub-contractor (see clause 3.17 of SBCSub/D/C).
- Single-phase supply of electricity at 110 V for tools and temporary lighting: All the contractor is required to do is provide a point of supply of 110 V electricity for the sub-contractor to connect to. Distribution of the electricity to the area of work, etc. is the responsibility of the sub-contractor.

  It is preferable for the location of the points of electricity supply to be agreed in advance and for that agreement to be recorded on a numbered document referred to in sub-contract particulars item 17 of SBCSub/D/A.
- Water supplied at the points identified in the sub-contract documents for temporary site accommodation and for the carrying out of the sub-contract works: The comments under the previous item apply equally here.
- Use of mess rooms: That is usually a site canteen.
- Use of sanitary accommodation: That is site toilets.
- Use of welfare facilities: For example, drying rooms, shower rooms and/or first-aid rooms.
- All reasonable non-exclusive use of hoisting facilities that the contractor has on site at the time the sub-contract works are being carried out: A sub-contractor should particularly note that:
  he or she only has reasonable and *non-exclusive* use of hoisting facilities;
  o he or she has no control over the type of hoisting facilities that will be provided by the contractor.

  For this reason it is usually sensible to agree with the contractor:

  o a hoisting/lifting register on a daily or weekly basis permitting the time periods for any hoisting required by the sub-contractor so that it can be planned in advance;
  o the hoisting facilities that will be provided (agreed in advance and recorded on a numbered document within sub-contract particulars item 17 of SBCSub/D/A).
- The benefit of all reasonable watching to be provided by the contractor under the main contract: The fact that watching may be provided by the contractor under the main contract does not absolve the sub-contractor from his or her liability for the loss or damage of his or her unfixed site materials.
- The provision of reasonable measures to prevent access by unauthorised persons: What will be reasonable will depend on each particular site. Further necessary consideration will also need to be given to the Work at Height Regulations 2005.

This list in item 6.1 of the sub-contract particulars is to be amended (as appropriate) and added to (if appropriate).

If there is insufficient room on SBCSub/D/A for further attendance items to be listed, further sheets should be used and should be annexed to the SBCSub/D/A as a numbered document.

It is important that the list of attendance items under item 6.1 is carefully considered and correctly amended, because all attendances (other than those finally listed) are to be provided by the sub-contractor.

7.6.1.3 What about other attendance items not listed in SBCSub/D/A?

In addition to the attendance items listed in item 6.1 of the sub-contract particulars, the following are some of the more common attendance items that the contractor and the sub-contractor may wish to discuss and/or agree are to be provided by the contractor; with any such agreement being recorded on a numbered document referred to in sub-contract particulars item 17 of SBCSub-/D/A. This is not intended to be a conclusive list, and the particular items of attendance will vary according to the particular circumstances:

- Responsibility for setting out of the sub-contract works: Commonly the contractor provides main grid lines and site datum levels from which the sub-contractor sets out the sub-contract works.
- Unloading/Distribution of materials: These are normally the responsibility of the sub-contractor. However, in certain circumstances (e.g. materials being delivered before a sub-contractor has a presence on site), it may be agreed that the contractor will unload (but will accept no liability for) those particular materials or goods, or distribute the materials to (or near to) the area of work.
- Space for storage of materials on site
- Covered/Secure store
- Personal Protective Equipment (PPE): This is normally the responsibility of the sub-contractor to ensure that his or her operatives have adequate and suitable PPE. However, if very special PPE is required, the contractor may agree to provide this.
- Protection of sub-contractor's goods, materials or works that are not fully and finally incorporated into the main contract works: This is normally the responsibility of the sub-contractor. Accordingly, if the contractor agrees to provide any protection materials or carry out any protection work, this needs to be confirmed as an additional attendance item.
- Responsibility for the control of or removal of water: This problem can often be overlooked, but should be discussed and agreed to avoid any future disagreements. This can be a particular problem in situations such as:
  o diamond drilling (where water is used to cool the cutting tool, etc.), particularly where the diamond drilling is carried out in an otherwise finished area of the works;
  o grit or sand blasting (with a water base), where the excess water needs to be removed (without the grit or sand blocking up the drain runs);
  o temporary roofs, where consideration is often not given to the discharge of rainwater into gutters, rainwater down-pipes and drains.
- Materials, fuel, power and/or water for commissioning and testing

Item 6.1 of the sub-contract particulars specifically notes that materials, fuel, power and/or water for commissioning and testing are excluded as an attendance item that a contractor would normally be expected to provide to a sub-contractor free of charge.

The sub-contractor may therefore need to allow for the necessary materials, fuel, power and/or water for commissioning and testing within his or her tender.

A point that is sometimes overlooked is that three-phase electricity is sometimes required for testing and commissioning (e.g. for lift installations, etc.) *before* the permanent supply has been connected. This factor can have major programme implications.

7.6.2 Who can use the scaffolding on site?

As noted above, one of the free issue items of attendance provided by the contractor in item 6.1 of the SBCSub/D/A is the provision and erection of all necessary scaffolding or other access/work equipment for all work at height. The only exclusion stated is where the provision of it is 'necessary solely and exclusively for the purpose of carrying out the Sub-contract works'.

In addition to this 'free' attendance item, clause 3.16.2 of SBCSub/D/C notes that the contractor, the sub-contractor, the contractor's persons and the sub-contractor's persons in common with all others having a like right shall for the purposes of the main contract works (and only for the purposes of the main contract works) be entitled to use any erected scaffolding belonging to or provided by the contractor or the sub-contractor while it remains erected.

Clause 3.16.2 relates to:

- both scaffolding belonging to or provided by the contractor and scaffolding belonging to or provided by the sub-contractor;
- the use of the scaffolding in question while it remains erected.

This presumably means that the scaffold may be used whilst it remains erected for its primary purpose, but that it will not remain erected thereafter for any secondary purpose.

Also, the scaffolding would (by implication) need to be used without alteration (e.g. amending lift heights, strengthening for loadings).

Clause 3.16.2 of SBCSub/D/C also adds that the use of this scaffold by either party is on the express condition that no warranty or other liability on the part of the contractor or the contractor's persons or of the sub-contractor or the sub-contractor's persons, as appropriate, shall be created or implied under this sub-contract in regard to the fitness, condition or suitability of such scaffolding.

This is a very significant point since it is the responsibility of both the contractor and the sub-contractor to ensure that their operatives on site are working from safe scaffolding.

With this point in mind, it may be the case that the sub-contractor (or the contractor, as appropriate) before using the scaffolding referred to under clause 3.16.2 of SBCSub/D/C would need to satisfy himself or herself, at least, that the scaffolding:

- was erected, altered and dismantled (as appropriate) by competent persons;
- had been regularly inspected;
- was secured to the building or structure (if appropriate) in enough places to prevent collapse;
- was strong enough to allow for the work that was proposed to be carried out from it, and for the materials that were proposed to be stored on it.

7.6.3 What does the sub-contract say about possible misuse of scaffolding, etc.?

Clause 3.16.3 of SBCSub/D/C provides that the contractor and the sub-contractor shall not, and shall respectively ensure that the contractor's persons and the sub-contractor's persons do not, wrongfully use or interfere with the plant, ways, scaffolding, temporary works, appliances or other property belonging to or provided by others; also, that no party should infringe upon any statutory requirements.

However, clause 3.16.3 also makes it clear that the above points do not affect the rights of the contractor or of the sub-contractor to carry out their respective statutory duties and contractual obligations under the sub-contract or the main contract.

Any misuse of scaffolding or other property belonging to or provided by the contractor is covered by the indemnity given by the sub-contractor to the contractor under clause 2.5.2 of SBCSub/D/C.

7.6.4 Temporary buildings

Site accommodation would normally include for site offices, workshops, stores and sheds, etc. (but not for mess rooms, sanitary accommodation and welfare facilities, which are normally provided by the contractor).

Clause 3.17 of SBCSub/D/C makes it clear that, subject to clause 3.16.1, it is for the sub-contractor (at his or her own expense) to provide, erect, maintain, move and subsequently remove all temporary site accommodation that he or she requires.

It may also be necessary for the sub-contractor to move his or her temporary site accommodation to a different location on site during the course of the works.

7.6.4.1 Who determines the location of sub contractor's site accommodation?

Clause 3.17 states that this is determined by the contractor, but also gives the sub-contractor the right to make a reasonable objection if, for example, the sub-contractor considers his or her site accommodation will be placed in an unsuitable location on site.

Normally, space on site is fairly limited, and it is therefore usual for the contractor to agree with the sub-contractor upon the number of pieces of individual site accommodation that will be permitted, the size of the various items of the site accommodation and any stacking arrangements that will be necessary.

Clause 3.17 of SBCSub/D/C states that for the purposes of the provision of, the erection of, the maintenance of, the moving of and/or the removal of temporary site accommodation by the sub-contractor, the contractor agrees to give all reasonable facilities.

7.6.4.2   *What are reasonable facilities?*

It is submitted that in the context of the above, all reasonable facilities simply means that the contractor will allow the sub-contractor, or give the opportunity to the sub-contractor, to provide, erect, maintain, move and/or remove temporary site accommodation; it does not mean that the contractor will provide any equipment or resources (e.g. cranage or labour) to the sub-contractor for any of the above-listed operations.

7.6.4.3   *Joint Fire Code*

Item 6.3 of the sub-contract particulars may have an impact on the construction of the temporary accommodation where the Joint Fire Code applies.

7.6.5   Site clearance – how is the sub-contractor's waste dealt with?

In respect of item 6.4 of the sub-contract particulars:

- if no entry is made, the normal (default) position is that all rubbish resulting from the carrying out of the sub-contract works is to be disposed of offsite by the sub-contractor;
- item 6.4. is only completed if there are any specific requirements for the manner in which rubbish is to be disposed of; for example, that the sub-contractor is to dispose of his or her rubbish to a central skip to be provided by the contractor; or may be that certain waste must be disposed of in a particular manner.

To the extent that the sub-contractor has liability for site clearance (as set out above), clause 3.18 of SBCSub/D/C states that the sub-contractor shall:

- clear away all rubbish resulting from his or her carrying out of the sub-contract works so as to keep access to the sub-contract works clear at all times;
- upon practical completion of the sub-contract works or of such works in each Section, clear up and leave those works clean and tidy to the reasonable satisfaction of the contractor, together with all areas made available to him or her (and used by him or her) for the purposes of carrying out the sub-contract works.

7.7 Health and safety and CDM

7.7.1 Health and safety

Clause 3.19 of SBCSub/D/C requires the sub-contractor to comply at no cost to the contractor with:

- all health and safety legislation relevant to the sub-contract works and the manner in which they are being carried out;
- all reasonable directions of the contractor to the extent necessary for compliance by the contractor and the sub-contractor with such legislation as it affects the sub-contract works;
- within the time reasonably required, any written request by the contractor for information reasonably necessary to demonstrate compliance by the sub-contractor with clause 3.19.

7.7.2 What is the relevant health and safety legislation?

Generally, the laws which govern health and safety are not industry specific, but they do relate to many construction activities (including design, where appropriate). There are several Acts of Parliament and regulations that apply.

The fundamental Act governing health and safety in construction is the Health and Safety at Work Act 1974. This Act has over 60 regulations, but the principal regulations of this Act which affect design and construction are:

- Health and Safety (First Aid) Regulations 1981
- Noise at Work Regulations 2005
- Electricity at Work Regulations 1989
- The Health and Safety Information for Employees Regulations 1989
- Manual Handling Operations Regulations 1992
- Personal Protective Equipment Regulations 1992
- Gas Safety (Installation and Use) Regulations 1994
- Reporting of Injuries, Diseases and Dangerous Occurrences Regulations 2013 (known as RIDDOR)
- Provision and Use of Work Equipment Regulations 1998 (known as PUWER 98)
- The Lifting Operations and Lifting Equipment Regulations 1998 (known as LOLER 98)
- The Control of Major Accident Hazards Regulations 1999 (known as COMAH)
- Management of Health and Safety at Work Regulations 1999
- The Chemicals (Hazards Information and Packaging for Supply) Regulation 2009 (known as CHIP). CHIP is due to be replaced by the European CLP Regulation and will therefore be revoked from 1 June 2015.
- The Control of Substances Hazardous to Health Regulations 2002 (known as COSHH)

- The Work at Height Regulations 2005 as amended by the Work at Height (Amendment) Regulations 2007
- The Construction (Design and Management) Regulations 2015 (known as the CDM Regulations).

Some of the more important elements of the above Acts and regulations are discussed at sections 7.7.2.1 to 7.7.2.20 of this Chapter below.

7.7.2.1   What was the intention of the Health and Safety at Work Act 1974?

The intention of this Act was to make further provision for securing the health, safety and welfare of persons at work; for protecting others against risks to health or safety in connection with the activities of persons at work; for controlling the keeping and use and preventing the unlawful acquisition, possession and use of dangerous substances; and for controlling certain emissions into the atmosphere.

The first part of section 2 of the Act contains a general statement of the duties of employers (which would include contractors and sub-contractors) to their own employees while at work and is qualified in subsection (2), which instances particular obligations to:

1. provide and maintain plant and systems of work that are safe and without risks to health. Plant covers any machinery, equipment or appliances, including portable power tools and hand tools.
2. ensure that the use, handling, storage and transport of articles and substances is safe and without risk.
3. provide such information, instruction, training and supervision to ensure that employees can carry out their jobs safely.
4. ensure that any workshop under his or her control is safe and healthy, and that proper means of access and egress are maintained, particularly in respect of high standards of housekeeping, cleanliness, disposal of rubbish and the stacking of goods in the proper place.
5. keep the workplace environment safe and healthy so that the atmosphere is such as not to give rise to poisoning, gassing or the encouragement of the development of diseases. Adequate welfare facilities should be provided.

In respect of the above, and in all of the following regulations, an employer would be a contractor and/or a sub-contractor, if and as appropriate.

Further duties are placed on an employer by:

- section 2(3) of the Act, which requires the employer to prepare and keep up to date a written safety policy supported by information on the organisation and arrangements for carrying out the policy. The safety policy has to be brought to the notice of employees; however, where there are five or less employees, this section does not apply.

    The safety policy should consist of three parts:
    - A general statement of intent
    - Details of the organisation (people and their duties)

○   Details of the practical arrangements (i.e. systems and procedures), e.g.:
  — safety training
  — safe systems of work
  — environmental control
  — safe place of work
  — safe plant and equipment
  — noise control
  — use of toxic materials
  — utilisation of safety committee(s) and safety representatives
  — fire safety and prevention
  — medical facilities and welfare
  — maintenance of records
  — accident reporting and investigation
  — emergency procedures

■   Section 2(6) of the Act, which requires an employer to consult with any safety representatives appointed by recognised trade unions to enlist their cooperation in establishing and maintaining high standards of safety.

■   Section 2(7) of the Act, which requires an employer to establish a safety committee if requested by two or more safety representatives.

The employer is not allowed by section 9 of the Act to charge any employee for anything done or provided to meet statutory requirements.

The employees' duties are laid down in sections 7 and 8 of the Act, and these sections state that, whilst at work, every employee must take care of the health and safety of himself or herself and of other persons who may be affected by his or her acts or omissions.

Also, employees should cooperate with their employer to meet legal obligations, and they must not, either intentionally or recklessly, interfere with or misuse anything, whether plant equipment or methods of work, provided by their employer to meet the obligations under this or any other related Act.

7.7.2.2   *Health and Safety (First Aid) Regulations 1981*

These regulations deal with the requirements for first aid.

7.7.2.3   *Noise at Work Regulations 2005*

These regulations require employers to take action to protect employees from hearing damage.

7.7.2.4   *Electricity at Work Regulations 1989*

These regulations require people in control of electrical systems to ensure that they are safe to use and are maintained in a safe condition.

7.7.2.5 The Health and Safety Information for Employees Regulations 1989

These regulations require (amongst other things) that employers display a poster telling employees what they need to know about health and safety.

7.7.2.6 Manual Handling Operations Regulations 1992

These regulations place an obligation on employers to carry out an assessment of manual handling activities undertaken and to reduce any identified risks. Manual handling includes lifting, pulling, pushing, carrying, lowering and turning.

Once the risks have been identified, the appropriate control measures must be identified and used.

7.7.2.7 Personal Protective Equipment Regulations 1992

These regulations place an obligation on employers to assess and review the provision and suitability of personal protective equipment at work; as noted above personal protective equipment is usually simply referred to as 'PPE'.

An employer must assess the need to provide PPE, but PPE should be provided only when other control measures have been examined and either implemented or dismissed as not being reasonably practicable.

It is the responsibility of an employer to provide PPE and the employer should not charge its employees for the PPE issued.

7.7.2.8 Gas Safety (Installation and Use) Regulations 1994

These regulations cover safe installation, maintenance, and use of gas systems and appliances in domestic and commercial premises.

7.7.2.9 Reporting of Injuries, Diseases and Dangerous Occurrences Regulations 2013 (known as RIDDOR)

These regulations require that employers and other persons who are in charge of workplaces, are to report and keep records of certain accidents, which must be reported to the Health and Safety Executive (usually referred to as the HSE).

These accidents are:

- a work-related accident that causes a person's death;
- a work-related accident that is a serious reportable injury (i.e. it comes within one of the listed categories);
- diagnosed cases of certain industrial diseases (i.e. as per the categories of work-related illness);
- certain types of 'dangerous occurrences' – these are various listed 'near-miss' incidents that have the potential to cause harm.

7.7.2.10 Provision and Use of Work Equipment Regulations 1998 (known as PUWER 98)

These regulations place an obligation on all employers to carry out an assessment of work equipment being used within a business and to reduce risks, so far as is reasonably practicable.

7.7.2.11 The Lifting Operations and Lifting Equipment Regulations 1998 (known as LOLER 98)

These regulations require that lifting equipment provided for use at work is strong and stable enough for the particular use, that it is positioned and installed to minimise any risks, that it is used safely and is subject to ongoing inspection.

7.7.2.12 The Control of Major Accident Hazards Regulations 1999 (known as COMAH)

These regulations require those who manufacture, store or transport dangerous chemicals or explosives in certain quantities to notify the relevant authority.

7.7.2.13 Management of Health and Safety at Work Regulations 1999

Regulation 3 of the Management of Health and Safety at Work Regulations 1999 places a legal duty on employers to carry out a risk assessment.

7.7.2.14 What is a risk assessment?

This is simply a careful examination of what, in the works to be carried out, could cause harm to people (including employees, other operatives and the public).

There are five basic steps to producing a risk assessment:

- Look for the hazards.
- Decide who might be harmed, and how they might be harmed.
- Evaluate the risks arising from the hazards identified and decide whether existing precautions are adequate or if more should be done.
- Record the findings.
- Review the assessment from time to time and revise it if necessary.

7.7.2.15 What is a method statement?

This is a method of control, usually used after a risk assessment of an operation is carried out.

The method statement is used to control the operation and to ensure that all concerned are aware of the hazards associated with the work and the safety precautions to be taken.

Method statements should be 'project specific', and one of the main criticisms of sub-contractors (in particular) is that they attempt to use 'generic' method statements (rather than 'project-specific' method statements).

7.7.2.16   The Chemicals (Hazards Information and Packaging for Supply) Regulation 2002 (known as CHIP)

This regulation requires suppliers to classify, label and package dangerous chemicals and provide safety data sheets for them.

7.7.2.17   The Control of Substances Hazardous to Health Regulations 2002 (known as COSHH)

These regulations require employers to control exposure to hazardous substances to prevent illness; both employees and others who may be exposed must be protected.

These hazardous substances include things such as adhesives, paints, cleaning agents, fumes from soldering and welding, biological agents and carcinogenic chemicals.

To comply with COSHH, there are eight basic steps to follow:

- Assess the risks.
- Decide what precautions are needed.
- Prevent or adequately control exposure.
- Ensure that control measures are used and maintained.
- Monitor the exposure.
- Carry out appropriate health surveillance.
- Prepare plans and procedures to deal with accidents, incidents and emergencies.
- Ensure employees are properly informed, trained and supervised.

7.7.2.18   The Work at Height Regulations 2005, as amended by the Work at Height (Amendment) Regulations 2007

These regulations came into force on 6 April 2005. Statistics record that nearly 25% of deaths at work each year are caused by falls from height. These regulations apply to all work at height where there is a risk of a fall liable to cause personal injury and are aimed at preventing deaths and injuries caused each year by falls at work by placing duties on employers, the self-employed and any person who controls the work of others.

7.7.2.19   The Construction (Design and Management) Regulations 2015 (known as the CDM Regulations)

These regulations are dealt with in more detail in the next section of this chapter.

7.7.2.20   *Where recorded as being applicable, how does schedule 1 (paragraph 2) of the Supplemental Sub contract Provisions seek to improve site health and safety?*

The Health and Safety provisions provided for in clause 3.19 of SBCSub/D/C are (where paragraph 2 is stated to apply in the sub-contract particulars) supplemented by schedule 1, paragraph 2, of the Supplemental Sub-contract Provisions.

This schedule expands upon the express requirements in clauses 3.19 and 3.20 requiring compliance with Health and Safety legislation, adding further layers of obligations in respect of same.

Paragraph 2.1 sets the scene and commences this Supplemental Sub-contract Provision by stating that 'without limiting either Party's statutory and/or regulatory duties and responsibilities and/or specific health and safety requirements of this sub-contract, the parties will endeavor to establish and maintain a culture and working environment in which health and safety is of paramount concern to everyone on the project'.

In addition to the specific health and safety provisions included in the Sub-contract, paragraph 2.2 states that the sub-contractor:

- further undertakes to ensure compliance with all and any approved codes of practice, which are either produced or promulgated by the HSE and/or Health and Safety Commission-approved codes.
- further undertakes to ensure that the sub-contractor's personnel and his or her supply chain members on site receive:
  - site-specific health and safety inductions;
  - regular health and safety refresher training.
- further undertakes that in respect of the personnel noted above:
  - they will, at all times, be afforded access to competent health and safety advice in accordance with regulation 7 of the Management of Health and Safety at Work Regulations 1999.
  - they receive a full and proper health and safety consultation in accordance with the Health and Safety (consultation with Employees) Regulations 1996.

7.8   The CDM Regulations (i.e. the Construction (Design and Management) Regulations 2015)

Obviously, all of the sub-contracts considered in this book address CDM 2015, which came into force on 6 April 2015.

Clause 3.20 of SBCSub/D/C deals with CDM Regulations as they apply to the sub-contractor.

To reflect their importance, clause 3.20 of SBCSub/D/C includes an express cross-undertaking by the parties to comply with CDM 2015 in relation to the main contract works and the site.

The main contract particulars annexed to the schedule of information in SBCSub/D/C will identify whether or not the project is notifiable.

7.8.1 Does the sub-contractor get paid for complying with CDM 2015?

No. Clause 3.20.3 of SBCSub/D/C states that the sub-contractor shall comply at no cost to the employer or the contractor with all reasonable requirements of the Principal Contractor relating to compliance by the sub-contractor with the CDM Regulations; it further notes that no extension of time shall be given for such compliance.

7.8.2 What is the purpose of the CDM 2015 Regulations?

The CDM Regulations require that health and safety is taken into account and managed throughout all stages of a project, from conception, design and planning, through to site work and subsequent to maintenance and repair of the building.

Accordingly, the CDM 2015 Regulations place health, safety and welfare duties upon everyone who takes part in the construction process—the employer, the designers, the contractors and the sub-contractors.

Despite certain similarities, the CDM 2015 Regulations are a more coherent and wider set of duties than its predecessor (the CDM 1994 Regulations), with increased emphasis on improving coordination and cooperation between the various parties, encouraging team working, reducing risk in the construction process and avoiding unnecessary bureaucracy.

All duty holders will need to ensure their health and safety procedures comply with the CDM 2015 Regulations.

7.8.3 What are the sub-contractors' duties under the CDM Regulations?

A sub-contractor's duties are the same as the contractor's. Additionally, if the sub-contractor designs, he or she assumes the obligations CDM 2015 places on designers.

7.8.4 When do the CDM Regulations apply?

The CDM Regulations apply to most common building, civil engineering and engineering construction work, and are defined widely in the regulations, including, for example, not only construction but also commissioning, renovation, repair, redecoration, demolition, etc.

The CDM Regulations apply to all design work carried out for construction purposes (including demolition and dismantling).

7.8.5 When must a project be notified to the Health and Safety Executive (HSE)?

Under CDM 2015, all projects fall into one of two categories: notifiable or non-notifiable. A project must be notified to the HSE if it lasts longer than 30 days and has more than 20 operatives working simultaneously on the it, or 500 man days of work.

7.8.6   What are the employer's duties under CDM 2015?

Under CDM 2015, employers (termed 'clients') can no longer delegate their duties to an agent (as occurred under CDM 1994).

The employer's main duties are described as follows:

■ Appoint a Principal Designer and a Principal Contractor for each project.
■ Take reasonable steps to ensure that the Principal Designer, Principal Contractor, project designers and any contractors they appoint directly are competent and adequately resourced.
■ Provide all pre-construction information in their possession to designers and contractors, including notification to the contractor of the minimum notice that will be given for the contractor's commencement of the works.
■ Provide the Principal Designer with pre-construction information, plus all health and safety information relating to the project in the employer's possession (or reasonably obtainable) as is likely to be needed for the Health and Safety File.
■ Ensure that the construction phase does not start until a suitable Construction Phase Plan has been prepared by the Principal Contractor and suitable welfare facilities are in place.
■ Ensure suitable project management arrangements for health and safety are in place (including the allocation of sufficient and other resources). Take reasonable steps to ensure that such arrangements are maintained and reviewed throughout the duration of the project.
■ After the Construction Phase Plan has been issued, take reasonable steps to provide access to the Health and Safety File and update this document, as necessary, with new information. In practical terms, the client must ensure that the Health and Safety File is reviewed and/or updated as necessary so that it is accurate at the completion of the construction work. This completed file must be kept available for others, that is, a new owner or for future construction work.

The reader is referred to the CDM 2015 Regulations for a full description of the duties placed upon clients.

7.8.7   What are the designer's duties under CDM 2015?

The term 'designer' includes everyone preparing drawings and specifications for the project, including variations and temporary works designs, and also includes producing detailed design work to satisfy a performance specification. Where appropriate, designers include architects, consulting engineers (e.g. structural engineers, building services engineers), contractors and subcontractors.

The designer's duties (which, in certain circumstances, could include a subcontractor) are, amongst other matters, as outlined next:

■ Ensure that the employer is aware of his or her duties under CDM 2015 before commencing work.

- Avoid foreseeable risks to the health and safety of any person when preparing or modifying the design by eliminating hazards where feasible, or reducing risks from those hazards that cannot be eliminated. The regulations oblige designers to give priority to collective measures in achieving this.
- Take reasonable steps to provide adequate information about all aspects of the design, construction or maintenance of the structure as will adequately assist the employer, other designers and the contractor in discharging their duties under the regulations.
- Ensure a Principal Designer is appointed before commencing work.
- Provide adequate information about all aspects of the design, construction or maintenance of the structure to the Principal Designer.
- Cooperate with the Principal Designer and any other designer involved in the project.
- Provide design information for inclusion in the Construction Phase Plan and the Health and Safety File.

The reader is referred to the CDM 2015 Regulations for a full description of the duties placed upon designers.

7.8.8   What are the Principal Designer's duties under CDM 2015?

The role of the Principal Designer includes, amongst other matters, for him or her to:

- advise and assist the employer in complying with his or her CDM 2015 duties.
- ensure the coordination and cooperation of health and safety between the relevant parties during the design and planning phase of the project.
- collect and distribute information, such as pre-construction information.
- liaise with the principal contractor regarding the health and safety file, the information the principal contractor needs to prepare the Construction Phase Plan and the ongoing design.
- take all reasonable steps to ensure that designers comply with their duties under the regulations.
- give advice about health and safety competence and resources needed for the project.
- ensure notice of the project is given to the HSE.
- prepare and update the Health and Safety File and pass the file to the employer on completion of the project (or, in some cases, pass to the Principal Contractor for completion).

The identity of the Principal Designer is normally provided in the main contract particulars as annexed to the schedule of information attached to the sub-contract, for example, SBCSub/D/A, DBSub/A, ICSub/A, etc.

The reader is referred to the CDM 2015 Regulations for a full description of the duties placed upon Principal Designers.

7.8.9 What are the Principal Contractor's duties under CDM 2015?

The Principal Contractor is, amongst other matters, to:

- plan, manage and monitor health and safety during the construction phase of the project, in liaison with other contractors, and ensure necessary cooperation and coordination between the relevant parties.
- develop a Construction Phase Plan before work starts on site. Thereafter, review and revise the Construction Phase Plan from time to time as appropriate. The plan may need to be developed during the construction phase to take account of changing conditions on site as work progresses or as the design changes. The Principal Contractor's Construction Phase Plan should take account of general issues, specific hazards, risk control measures and the general principles of risk assessment.
- provide the Principal Designer with all information he or she requires for inclusion in the Health and Safety File.
- ensure suitable welfare facilities are provided from the start and maintained throughout the construction phase.
- ensure all contractors are provided with relevant health and safety information, and that all workers have site inductions and (as necessary) further information and training.

It should be noted that whilst the contractor would normally be the Principal Contractor, this may not always be the case.

The identity of the Principal Contractor will be provided in the main contract particulars as annexed to the schedule of information to the sub-contract.

The reader is referred to the CDM 2015 Regulations for a full description of the duties placed upon the Principal Contractor.

7.8.10 What are the contractor's (and sub-contractor's) duties under CDM 2015?

The contractor's (and sub-contractor's) duties include, amongst other matters, to:

- ensure that the employer is aware of his or her duties under CDM 2015 before commencing work.
- plan, manage and monitor his or her work and that by others on his or her behalf to ensure, as far as is reasonably practicable, that his or her works are executed without risks to health and safety.
- ensure that they and anyone they employ or engage (e.g. sub-contractors, workers) are competent.
- ensure any sub-contractor used by the contractor is notified of the minimum amount of time that will be allowed for his or her preparation and planning prior to commencement on site.
- ensure his or her workers (whether employed or self-employed and undertaking work under the contractor's control) are provided with necessary information and training to permit work to be carried out safely.

- cooperate and coordinate their work with others as required.
- ensure the provisions of adequate welfare facilities.
- ensure a Principal Designer is appointed and the HSE notified before commencing work.
- cooperate with the Principal Designer.
- cooperate with the Principal Contractor, including providing information which may affect the health and safety of those executing the construction work or any person affected by it.
- comply with the Principal Contractor's reasonable directions, site rules and the Construction Phase Plan.
- notify the Principal Contractor of any matters that may require a review, alteration or addition of the construction phase plan and/or problems with the plan or risks identified during their work that have significant implications for the management of the project.
- issue the Principal Contractor with information for the Health and Safety File.
- notify the Principal Contractor of any death, accidents or dangerous occurrences.
- provide adequate information about all aspects of the design, construction or maintenance of the structure to the Principal Designer and/or the Principal Contractor.

The sub-contractor's duties are the same as the contractors. The CDM 2015 Regulations place extensive duties on contractors (and sub-contractors) regarding health and safety and welfare. The reader is referred to the CDM 2015 Regulations for a full description of the duties placed upon contractors (and sub-contractors).

7.8.11   How does the role of the Principal Contractor affect the sub-contractor?

Sub-contractors are required to help the Principal Contractor to achieve safe and healthy site conditions. They should cooperate with other sub-contractors working on the site and provide health and safety information (including risk assessments) to the Principal Contractor.

7.8.12   How does health and safety competence affect sub-letting?

The CDM Regulations require that when sub-letting, a contractor should satisfy himself or herself that those who are to do the work are:

- competent in relevant health and safety issues;
- intend to allocate adequate resources, including time, equipment and properly trained workers, to do the job safely and without risks to health.

It is quite common for sub-contractors to be asked to complete a health and safety competence questionnaire before their tender is considered for acceptance.

Additionally, to reinforce the importance of CDM 2015, express terms are also usually included in the sub-subcontract. For example, clause 3.3.2.1 of the SBCSub/D/C makes it a requirement that a sub-subcontract will include an express undertaking to comply with CDM 2015 in relation to the main contract works at the site.

7.8.13   What is pre-construction information?

The pre-construction information is provided by the employer to designers and contractors who are, or may be, appointed. The following information is to be included:

■   any information about and/or affecting the site and/or construction work;
■   any information concerning the proposed use of the structure as a workplace;
■   the minimum amount of time before the construction phase allowed to contractors for planning and preparation for construction work;
■   any information in any existing Health and Safety File.

Its purpose is to ensure that information relevant to health and safety is passed on to those who need it so that significant risks during the work can be anticipated and planned for.

7.8.14   What is the Construction Phase Plan?

The Construction Phase Plan (previously known as the Health and Safety Plan) is developed by the Principal Contractor to address key issues of health, safety and welfare relevant to the project.
Issues which need to be considered for inclusion in the plan include:

■   How health and safety will be managed during the construction phase.
■   What high-risk activities will require risk assessments and method statements to be produced?
■   Information about welfare arrangements
■   Information on necessary levels of health and safety training for those working on the project and arrangements for project-specific awareness training and refresher training such as toolbox talks: A toolbox talk is a method of passing health and safety information on to the site operatives at site level. The toolbox talk is normally given by the site safety representative to a group of operatives on site, and is normally in relation to a particular health and safety issue (e.g. the safe use of electric power tools or the need for leading edge protection).
■   Arrangements for monitoring compliance with health and safety law.

The Construction Phase Plan should be developed before construction work actually starts, and should then be reviewed as necessary to account for changing project circumstances.
Clause 3.20 of SBCSub/D/C obliges the contractor to ensure that the sub-contractor is supplied forthwith with a copy of any development of the Construction Phase Plan by the Principal Contractor.

7.8.15   What is the Health and Safety File?

The Health and Safety File is a record of information for the employer or the end user.

The Principal Designer/Principal Contractor (as appropriate) ensures that it is produced at the end of the project and that it is then passed to the employer. The file gives details of health and safety risks that will have to be managed during maintenance, repair, renovation or demolition.

Contractors (and sub-contractors) should pass information that may affect the maintenance, repair, renovation or demolition of the finished building to the Principal Designer/Principal Contractor (as appropriate) for inclusion within the file.

The employer should make the file available to those who will work on any future design, construction, maintenance or demolition of the structure (and this may include the contractor or the sub-contractor when they are carrying out defect rectification work).

A Health and Safety File will often incorporate operating and maintenance manuals and as-built drawings (prepared by the contractor and the sub-contractors), as well as general details of the construction methods and materials used.

For example, clause 3.20 of SBCSub/D/C requires the sub-contractor to provide to the contractor, within the time reasonably required by the contractor, such information in respect of the sub-contract works as is reasonably necessary to enable the contractor to comply with clause 3.23.4 of the main contract conditions, which relates to the information reasonably required by the Principal Designer/Principal Contractor (as appropriate) for inclusion in the Health and Safety File.

7.9   Suspension of the main contract by the contractor

7.9.1   The General Position

All the JCT main contracts to which the sub-contracts in this book refer contain express rights for the main contractor to suspend the performance of any or all of his or her obligations under the main contract where the employer has failed to make payment in the specified manner. Such rights are of course subject to the contractor's strict compliance with the notice requirements and procedure stated in the main contract.

Naturally, suspension by the contractor of his her performance of any or all of his or her obligations under the main contract will raise questions by the sub-contractor as to what action, if any, he or she should take under the sub-contract.

In such circumstances, the sub-contractor should refer to the provisions of the sub-contract itself for the answer.

However, as a general principle, unless the contract provides otherwise (which the unamended JCT sub-contracts do not):

1. there is no automatic right for a sub-contractor to suspend his or her performance under the sub-contract in circumstances where the contractor has suspended any or all of his or her obligations under the main contract works.

2.   the sub-contractor should not stop carrying out the sub-contract works unless he or she is instructed or directed by the contactor to do so.

Of course, the above possibility may cause great difficulties in practice, for example, in respect of health and safety considerations, and this is a matter that would need to be considered by a sub-contractor on a project-by-project basis.

7.9.2   Does the sub-contract provide for the sub-contractor to be notified of the contractor's proposed suspension/suspension of his or her performance under the main contract?

The SBCSub/D deals with this in clause 3.21.

7.9.2.1   Position where the contractor intends to suspend

Clause 3.21 states that if the contractor under clause 4.14 of the main contract conditions gives the employer a written notice of his or her intention to suspend the performance of any or all of his or her obligations under the main contract because of non-payment by the employer, he or she is to immediately copy that notice to the sub-contractor.

7.9.2.2   Position where the contractor actually suspends the main contract

In circumstances where the contractor actually suspends the performance of any or all of his or her obligations under the main contract, clause 3.21 obliges the contractor to immediately notify the sub-contractor of this occurrence.

7.9.3   What should the sub-contractor do when notified that the contractor has suspended his or her performance under the main contract?

The sub-contractor *must not* suspend the performance of the sub-contract works simply because he or she has received notice from the contractor that he or she has suspended the performance of the main contract works.

Instead, the sub-contractor should, where possible continue to proceed regularly and diligently with his or her sub-contract works, until and/or unless he or she receives an express direction from the contractor to cease the carrying out of the sub-contract works.

If the sub-contractor were to suspend his or her performance in the absence of an express direction from the contractor, this may be found to be a wrongful suspension by the sub-contractor, and this may amount to a repudiation by the sub-contractor, which, if accepted by the contractor, may lead to the sub-contractor suffering all of the contractor's damages arising from that repudiation.

7.9.4   In what circumstances should the sub-contractor cease the sub-contract works when the main contract is suspended by the contractor?

7.9.4.1   Health and safety issues – all sub contracts

Surprisingly, all of the sub-contracts discussed in this book fail to address a real practical issue.

What happens if the suspension of the main contract renders the execution of the sub-contract works unsafe, that is, withdrawal of health and safety supervisor, scaffolding, welfare provisions, etc., but the contractor has not directed the sub-contractor to cease the carrying out of the sub-contract work?

Obviously, each case will vary on its own particular facts, but given the potential ramifications and statutory penalties where this applies, common sense dictates that the sub-contractor must immediately raise any health and safety concerns with the contractor and request his or her urgent response and action.

The provisions included in SBCSub/D for directions by the contractor are considered below.

7.9.4.2   Directions under the sub contract are received from the contractor to cease the sub contract works

Clause 3.22 of SBCSub/D/C deals with directions from the contractor to the sub-contractor in two parts:

1. Direction to cease sub-contract works, and further directions in respect of such cessation (clause 3.22.1 refers)
2. Direction to recommence the sub-contract works, and further directions in respect of such recommencement (clause 3.22.2 refers)

This is discussed below.

7.9.4.2.1   Clause 3.22.1: the contractor's direction to cease sub-contract works, and further directions in respect of such cessation

Following the contractor's suspension of any or all of his or her obligations under the main contract, clause 3.22.1 states that the contractor 'may' direct the sub-contractor to cease the carrying out of the sub-contract works and after that may issue to the sub-contractor such further directions as necessary.

The word 'may' denotes that the contractor is not obliged to issue a direction to the sub-contractor to cease the carrying out of the sub-contract works; it is entirely at the contractor's discretion to issue (or not issue) this direction.

There are potentially serious ramifications if the sub-contractor suspends without first receiving a contractor's direction under clause 3.22.1.

The sub-contractor should therefore ensure that he or she has first received a clear written direction to suspend works before suspending his or her works. Unless and until such a direction is received, the sub-contractor must proceed regularly and diligently with his or her sub-contract works.

If such a direction is issued, the contractor may issue further directions as may be necessary following such cessation. Although the SBCSub/D/C does not elaborate upon what those 'further directions' may be, it would be reasonable to assume that they would include for things such as a requirement that unfinished works are adequately protected, and that the site working area is left secure.

7.9.4.2.2   Clause 3.22.2: direction to recommence the sub-contract works, and further directions in respect of such recommencement (clause 3.22.2 refers)

This clause only applies in circumstances where the contractor has issued a direction to the sub-contractor to cease the sub-contract works pursuant to clause 3.22.1.

Clause 3.22.2 of SBCSub/D/C states that if the contractor resumes the performance of his or her obligations under the main contract, the contractor shall, if he or she has directed the sub-contractor to cease the carrying out of the sub-contract works, direct the sub-contractor to recommence those works and may issue further directions in regard to such recommencement.

Again, although the SBCSub/D/C does not elaborate upon what those 'further directions' may be, it would be reasonable to assume that they would allow for such things as the removal of protection from unfinished works, and the cleaning down of work left standing during the period of suspension to allow other trades to proceed.

The sub-contractor's position is safeguarded from any suspension due to the above in respect of:

■ Time – by the inclusion of clause 2.19.7 of SBCSub/D/C as a Relevant Sub-contract Event to be considered by the contractor when giving the sub-contractor an extension of time to the completion date or to a sectional completion date of the sub-contract works
■ Loss and expense – by the inclusion of clause 4.20.4 of SBCSub/D/C as a Relevant Sub-contract Matter to be considered by the contractor when agreeing the amount of direct loss and expense due to the sub-contractor

7.9.5   Other provisions – strikes

What happens when the main contract works are affected by strikes, etc.?

Clause 3.23 of SBCSub/D/C deals with strikes.

This clause says that if the main contract works are affected by a strike, lock-out or local combination of workmen affecting any of the trades employed upon them or any of the trades engaged in the preparation, manufacture or transportation of any of the goods or materials required for the main contract works:

■ neither the contractor nor the sub-contractor shall be entitled to make any claim upon the other for any loss and expense resulting from such action;

- the contractor shall take all reasonable steps to keep the site open and available for the use of the sub-contractor;
- the sub-contractor shall take all reasonable practical steps to continue with the sub-contract works.

Clause 3.23 further notes that nothing in clause 3.23 shall affect any other right of the contractor or the sub-contractor under the sub-contract if such action occurs.

Although not elaborated upon, this caveat would probably, for example, protect the parties' respective position in respect of the provisions for termination under the sub-contract. Clause 2.19.14 of SBCSub/D/C also safeguards the position of the sub-contractor by specifically including a relevant event for such matters.

7.10   Benefits under the main contract

Subject to the sub-contractor providing an appropriate indemnity and security for costs to accord with the reasonable requirements of the contractor, clause 3.24 of SBCSub/D/C grants the sub-contractor a right to require the contractor to take certain actions under the main contract which may be relevant in the case of contentious opinions and instructions by the employer under the main contract (provided that those actions are relevant to the sub-contract works and are not inconsistent with any other express terms of the sub-contract).

Any such action as outlined above is to be undertaken at the expense of the sub-contractor.

7.11   Certificates/statements or notices under the main contract

For obvious reasons the sub-contractor needs to be informed by the contractor of certain key main contract certificates or statements or notices (the precise terminology relates to the main contract-sub-contract relationship) which have a direct bearing on the sub-contract.

7.11.1   How does the sub-contractor receive this information?

Clause 3.25 of SBCSub/D/C deals with certain key statements and notices issued under the main contract, and how the sub-contractor goes about obtaining such information from the contractor.

Clause 3.25 obliges the contractor, upon receipt of a written request from the sub-contractor, to notify the sub-contractor of the dates that the following documents are issued under the main contract:

1. The fixing or confirmation of the completion date for the main contract works or any section: Comment—The completion date for the main contract works is relevant in respect of any liquidated damages which the contractor may be required to pay under the provisions of the main contract, which in certain circumstances the sub-contractor may be liable for.

2. Any written notice or certificate issued under clause 2.33 or 2.35 of the main contract conditions: Comment—Deals with partial possession by the Employer under the main contract:

   ■ Clause 2.33 of the main contract conditions: Deals with the notice that is given by the Architect/Contract Administrator to the contractor identifying the part or parts taken into possession by the Employer and the date, and is deemed to be practical completion of that portion of the main contract works.

   ■ Clause 2.35 of the main contract conditions: Deals with the certificate of making good issued by the Architect/Contract Administrator to the contractor for the relevant part taken into possession.

3. The practical completion certificate or each sectional completion certificate for the main contract works: Comment—The practical completion certificate date or each sectional completion certificate date in respect of the main contract works is the deemed latest date for the practical completion and/or the sectional completion of the sub-contract works.

4. Each notice of completion of making good (defects): Comment—The notice of completion of making good for the main contract works or relevant section establishes the latest possible date that triggers the release of the balance of any retention to the sub-contractor.

5. The final certificate: Comment—The final certificate under the main contract triggers the release of the final payment to be made by the contractor to the sub-contractor.

7.12 Equivalent sub-contract provisions

In the table below, the equivalent sub-contract provisions to that within the SBCSub/D/A or SBCSub/D/C in the text above are listed for the SBCSub/A or SBCSub/C, DBSub/A or DBSub/C, ICSub/A or ICSub/C and ICSub/D/A or ICSub/D/C.

SBCSub/D/A or SBCSub/D/C	SBCSub/A or SBCSub/C	DBSub/A or DBSub/C	ICSub/A or ICSub/C	ICSub/D/A or ICSub/D/C
Clause 2.13.1	N/A	Clause 2.13.1	N/A	Clause 2.21.1
Clause 3.1	Clause 3.1	Clause 3.1	Clause 3.1	Clause 3.1
Clause 3.2.1	Clause 3.2	Clause 3.2.1	Clause 3.2.1	Clause 3.2.1
Clause 3.2.2	N/A	Clause 3.2.2	N/A	Clause 3.2.2
Clause 3.3.1	Clause 3.3.1	Clause 3.3.1	Clause 3.3.1	Clause 3.3.1
Clause 3.3.2.1	Clause 3.3.2.1	Clause 3.3.2.1	Clause 3.3.2.1	Clause 3.3.2.1
Clause 3.3.2.2	Clause 3.3.2.2	Clause 3.3.2.2	Clause 3.3.2.2	Clause 3.3.2.2
Clause 3.8	Clause 3.8	Clause 3.8	Clause 3.7	Clause 3.7
Clause 3.9	Clause 3.9	Clause 3.9	Clause 3.8	Clause 3.8
Clause 3.10	Clause 3.10	Clause 3.10	Clause 3.9	Clause 3.9

(Continued)

SBCSub/D/A or SBCSub/D/C	SBCSub/A or SBCSub/C	DBSub/A or DBSub/C	ICSub/A or ICSub/C	ICSub/D/A or ICSub/D/C
Clauses 3.11.1-Clause 3.11.3	Clauses 3.11.1-3.11.3	Clauses 3.11.1-3.11.3 Note that SBCSub/D/C's reference to directions in respect of materials or goods or workmanship, where the approval of the quality and standards is a matter for the opinion of the architect/contract administrator, is omitted in DBSub/C.	Clauses 3.10.1-3.10.3 However, ICSub/C deals with this area differently to SBCSub/D/C in its provisions and drafting.	Clauses 3.10.1-3.10.3 However, ICSub/D/C deals with this area differently to SBCSub/D/C in its provisions and drafting.
Clause 2.19.2.3	Clause 2.19.2.3	Clause 2.19.2.3	Clause 2.13.2.3	Clause 2.13.2.3
Clause 4.20.2.3	Clause 4.20.2.3	Clause 4.20.2.3	Clause 4.17.2.3	Clause 4.17.2.3
Clause 3.12.1	Clause 3.12.1	Clause 3.12.1	N/A ICSub/C has no provision allowing non-complaint work to remain. See clauses 3.11.1 and 3.11.2.	N/A ICSub/D/C has no provision allowing non-complaint work to remain. See clauses 3.11.1 and 3.11.2.
Clause 3.12.2	Clause 3.12.2	Clause 3.12.2		
Clause 3.12.3	Clause 3.12.3	Clause 3.12.3		
Clauses 3.13.1 and 3.13.2	Clauses 3.13.1 and 3.13.2	Clauses 3.13.1 and 3.13.2	Clause 3.11.2 Permits the contractor to issue such directions as are reasonably necessary, and the sub-contractor is to comply without any cost to the contractor	Clause 3.11.2 Permits the contractor to issue such directions as are reasonably necessary, and the sub-contractor is to comply without any cost to the contractor
Clause 3.14.1	Clause 3.14.1	Clause 3.14.1	Clause 3.12.1	Clause 3.12.1
Clause 3.14.2	Clause 3.14.2	Clause 3.14.2	Clause 3.12.2	Clause 3.12.2
Clause 3.15	Clause 3.15	Clause 3.15	Clause 3.13	Clause 3.13
Schedule 5	Schedule 5	Schedule 5	N/A	N/A
Sub-contract particulars item 6	Sub-contract particulars item 6	Sub-contract particulars item 6	Sub-contract particulars item 6	Sub-contract particulars item 6
Clause 3.16.1	Clause 3.16.1	Clause 3.16.1	Clause 3.14.1	Clause 3.14.1
Clause 3.16.2	Clause 3.16.2	Clause 3.16.2	Clause 3.14.2	Clause 3.14.2
Clause 3.16.3	Clause 3.16.3	Clause 3.16.3	Clause 3.14.3	Clause 3.14.3

SBCSub/D/A or SBCSub/D/C	SBCSub/A or SBCSub/C	DBSub/A or DBSub/C	ICSub/A or ICSub/C	ICSub/D/A or ICSub/D/C
Clause 3.17	Clause 3.17	Clause 3.17	Clause 3.15	Clause 3.15
Clause 3.18.1	Clause 3.18.1	Clause 3.18.1	Clause 3.16.1	Clause 3.16.1
Clause 3.18.2	Clause 3.18.2	Clause 3.18.2	Clause 3.16.2	Clause 3.16.2
Clause 3.19.1	Clause 3.19.1	Clause 3.19.1	Clause 3.17.1	Clause 3.17.1
Clause 3.19.2	Clause 3.19.2	Clause 3.19.2	Clause 3.17.2	Clause 3.17.2
Clause 3.19.3	Clause 3.19.3	Clause 3.19.3	Clause 3.17.3	Clause 3.17.3
Clause 3.20.1	Clause 3.20.1	Clause 3.20.1	Clause 3.18.1	Clause 3.18.1
Clause 3.20.2	Clause 3.20.2	Clause 3.20.2	Clause 3.18.2	Clause 3.18.2
Clause 3.20.3	Clause 3.20.3	Clause 3.20.3	Clause 3.18.3	Clause 3.18.3
Clause 3.20.4	Clause 3.20.4	Clause 3.20.4	Clause 3.18.4	Clause 3.18.4
Clause 3.21	Clause 3.21	Clause 3.21	Clause 3.19	Clause 3.19
Clause 3.22.1	Clause 3.22.1	Clause 3.22.1	Clause 3.20.1	Clause 3.20.1
Clause 3.22.2	Clause 3.22.2	Clause 3.22.2	Clause 3.20.2	Clause 3.20.2
Clause 2.19.6	Clause 2.19.6	Clause 2.19.6	Clause 2.13.6	Clause 2.13.6
Clause 4.20.4	Clause 4.20.4	Clause 4.20.4	Clause 4.17.3	Clause 4.17.3
Clause 3.23	Clause 3.23	Clause 3.23	Clause 3.21	Clause 3.21
Clause 2.19.13	Clause 2.19.13	Clause 2.19.13	Clause 2.13.13	Clause 2.13.13
Clause 3.24	Clause 3.24	Clause 3.24	Clause 3.22	Clause 3.22
Clause 3.25	Clause 3.25	Clause 3.25	Clause 3.23	Clause 3.23

8 Payment

8.1 Introduction

The right to payment (both interim and final) is obviously a matter of paramount importance to all parties in the construction industry. The issue of non-payment and/or under-payment is often a great source of dispute between contractors and sub-contractors. The old 'Construction Act'[1] and the new 'Construction Act'[2] both made the right to interim payments a statutory requirement. The two Acts dealt with the issues in slightly different ways, and the sub-contracts under consideration in this book are governed by the new 'Construction Act' rather than the old 'Construction Act'. The payment regime is now governed by Payment Notices, Default Payment Notices and Pay Less Notices, and it is now more important than ever that the payment provisions of the sub-contracts in question are followed meticulously.

Against this background, in this chapter, the following matters will be dealt with, using the sub-contract clause references, etc. contained in the SBCSub/D/A or SBCSub/D/C (as appropriate). Whilst it is clearly beyond the scope of this book to review every nuance of the other sub-contract forms under consideration, the equivalent provisions (where applicable) within the SBCSub/A or SBCSub/C, the DBSub/A or DBSub/C, the ICSub/A or ICSub/C and the ICSub/D/A or ICSub/D/C are given at the table at the end of this chapter. It must be emphasised that before considering a particularly issue, the actual terms of the appropriate edition of the relevant sub-contract should be reviewed by the reader (and/or legal advice should be sought as appropriate) before proceeding with any action/inaction in respect of the sub-contract in question.

In this chapter, the following matters are dealt with:

- The amount due in respect of interim payments
- Unfixed materials
- The amount due in respect of the final payment
- Retention
- Payment due dates and final dates for payment for interim payments

---

[1] The Housing Grants, Construction and Regeneration Act 1996
[2] The Local Democracy, Economic Development and Construction Act 2009

The JCT 2011 Building Sub-contracts, First Edition. Peter Barnes and Matthew Davies.
© 2016 John Wiley & Sons, Ltd. Published 2016 by John Wiley & Sons, Ltd.

- Payment Notices and Pay Less Notices for interim payments
- The payment due date and the final date for payment for the final payment
- Payment Notices and Pay Less Notices for the final payment
- VAT (Value Added Tax)
- Construction Industry Scheme (CIS)
- Interest
- Sub-contractor's right of suspension
- Fluctuations

8.2   The amount due in respect of interim payments

In line with clause 4.9.4, the amount due in interim payments is based either on:

- the value of any stage payments agreed between the contractor and the sub-contractor (e.g. payment at completion of elements of the work)
  or, if no stage payments have been so agreed, which in the vast majority of cases is the case:
- the gross valuation as referred to in clause 4.13.
  Irrespective of which of the above options is used, the following deductions are to be made:
  (1)   any amount which may be deducted and retained as retention in respect of the sub-contract works in accordance with clause 4.15 (i.e. retention monies);
  (2)   the sums previously due as interim payments under the sub-contract

The gross valuation as referred to in clause 4.13 is the total of the amounts referred to in clauses 4.13.1–4.13.8 (see the following list) less the total of the amounts referred to in clause 4.13.9, applied up to and including a date not more than 7 days before the date when the interim payment becomes or would become due.

This gross valuation would therefore include, in respect of clauses 4.13.1–4.13.8:

- the total of the sub-contract work on site properly executed by the sub-contractor, including any variation work executed, and, where applicable, with the adjustment for fluctuations under option C. Where a priced activity schedule is included in the numbered documents, the value is to be based upon the percentage that is completed of each activity in that activity schedule;
- the total value of the materials and goods delivered to or adjacent to the main contract, provided that those materials and goods are adequately protected against weather and other casualties, and provided that they have not been delivered prematurely;
- the total value of any off-site materials (if those materials are listed items under the sub-contract, and assuming that the pre-conditions for the payment of off-site materials have been met);
- the total value of payments made or costs incurred by the sub-contractor under clauses 2.10, 2.21 or 2.23 of the main contract conditions (i.e. in

respect of fees or charges legally demandable, and in respect of royalties and patent rights);
- any loss and expense payments (including those relating to the proper suspension of the works by the sub-contractor);
- the cost of the restoration of the sub-contract works and the replacement or repair of any sub-contract site materials that are lost or damaged, and the removal and disposal of debris, which under insurance clause 6.7.4 is to be treated as a variation;
- any amount payable to or deductible from the sub-contractor under fluctuations option A or B, if applicable;
- less; any amount deductible from the sub-contractor in respect of inaccurate setting out by the sub-contractor or to defects or other faults in the sub-contract works, and for any non-compliant work carried out by the sub-contractor which is allowed to remain in place (as clause 4.13.9).

8.3 Unfixed materials

8.3.1 Unfixed materials on site

It should be noted that even if payment is not made for any materials or goods delivered to or adjacent to the main contract works (which are intended for the main contract works) by the sub-contractor, these materials or goods shall not be removed from site unless the contractor consents in writing to such removal. This is confirmed under clause 2.15.1, which also makes it clear that such consent shall not be unreasonably delayed or withheld.

Clause 2.15.2 notes that when the value of any of the sub-contractor's unfixed materials or goods on site have been included in any interim payment under which the amount properly due to the contractor has been paid to him or her by the employer, those materials or goods shall be and shall become the property of the employer, and the sub-contractor shall not deny that they are and have become the property of the employer. Therefore, in such a scenario, the sub-contractor's unfixed materials and goods would become the property of the employer even though the sub-contractor may not have been paid for them.

Further, clause 2.15.3 notes that if the contractor pays the sub-contractor for any unfixed materials or goods on site, in advance of a value for such materials and goods being included in any interim payment (under the main contract), such materials or goods shall be and shall become the property of the contractor upon payment being made by him or her to the sub-contractor.

The above clauses are designed to protect the employer and the contractor from 'right of title' claims by the sub-contractor in respect of unfixed materials and goods on site.

The above concern only exists whilst the materials or goods are unfixed. Materials and goods which are to be incorporated into the works in almost all construction projects are bound at some stage to become 'attached to the soil', and thus become 'fixtures' to the land. Once the materials and goods become fixtures, they become in effect the property of the freeholder (normally the employer) in any event.

8.3.2   Unfixed materials off site

The first point is that payment for off-site materials will not be made unless the materials are listed items under the sub-contract particulars item 8.

However, even if the materials or goods in question are listed items, the sub-contractor is only entitled to be paid if the following conditions have been fulfilled:

(1)   The listed items are in accordance with the sub-contract. This effectively means that the item must be in line with the specification requirements. However, a listed item could also not be in accordance with the sub-contract if, because of a variation, the work requiring the listed item had been omitted or altered.

(2)   The sub-contractor has provided the contractor with reasonable proof that the property in such listed items is vested in the sub-contractor.

(3)   The listed items are insured for their full value against loss or damage under a policy of insurance protecting the rights of the employer, the contractor and the sub-contractor. The insurance needs to be in force from the period when the property (right of title) of the materials or goods passes to the sub-contractor up until the time that those materials or goods are delivered to, or adjacent to, the main contract works.

(4)   At the premises where the listed items have been manufactured or assembled or are stored, the materials or goods are set apart or have been clearly and visibly marked (individually or in sets) by letters or figures or by reference to a pre-determined code, and there is in relation to such items clear identification of the contractor and the employer to whose order they are held, and their destination as the main contract works.

(5)   That, if required in the sub-contract particulars, a bond in favour of the contractor or the employer (as the contractor directs) is provided from a surety approved by the contractor, in the amount set out in the sub-contract particulars and in the terms set out in schedule 3 Part 1 to the sub-contract. The form of bond in question is what is known as an 'on-demand' bond. An 'on-demand' bond is a bond that can be called upon without any condition (i.e. purely 'on demand'), and the sub-contractor needs to be aware of this fact before entering into the bond.

8.4   The amount due in respect of the final payment

Pursuant to clause 4.6.1 the sub-contractor is to send to the contractor all documents necessary for calculating the final sub-contract sum not later than 4 months after practical completion of the sub-contract works. In accordance with clause 4.6.2, the contractor shall not later than 8 months after receipt of the documents provided by the sub-contractor in line with clause 4.6.1, above, and in any event before the issue of the Final Certificate under the Main Contract, prepare and send to the sub-contractor a statement of the calculation of the final sub-contract sum.

The sub-contract can be placed on the basis of either a sub-contract sum (article 3A) or a sub-contract tender sum (article 3B):

- Using the sub-contract sum means that the final sub-contract sum will be calculated on an adjustment basis (as below).
- Using the sub-contract tender sum means that final sub-contract sum will be calculated on a re-measurement basis (as below).

8.4.1   Final sub-contract sum – adjustment basis (i.e. sub-contract based upon sub-contract sum)

The adjustment basis is generally referred to as a lump sum contract.

In this context, a lump sum contract is a contract to complete specified works for a lump sum price, but payment would still normally be paid if the specified works were not entirely complete. A lump sum contract is different to an 'entire' contract, which rarely applies in any event, where payment is only made if the contract works are *entirely* complete.

Under the adjustment basis, the quality and quantity of work included in the sub-contract sum is that set out in the bills of quantities and the sub-contract design documents, or where there are no bills of quantities, the quality and quantity of work is that included in the sub-contract documents taken together.

If the contractor provides quantities and these are contained in the sub-contract documents, then the quantity of work contained in the sub-contract sum shall be in line with those quantities.

When the adjustment basis applies, the sub-contract sum shall not be altered in any way other than in accordance with the express provisions of the conditions (e.g. by the issue of a variation instruction). This approach is in line with the traditional approach of a lump sum contract, with additions and/or omissions to that lump sum price for variations.

Importantly, it must be noted that other than for any errors or inadequacies in the bills of quantities, any error, whether an arithmetic error or otherwise, shall be deemed to have been accepted by the parties.

Therefore, if there is an arithmetical or other error in the computation of a bill of quantities item, a bill of quantities page total, a bill of quantities section total or a bill of quantities summary page, then these errors are deemed to have been accepted by the parties and no subsequent adjustment will be made for such errors.

Under the adjustment basis, the final sub-contract sum is calculated by making the following adjustments to the sub-contract sum:

(1) The amount stated (either an addition or an omission) by the contractor in his or her acceptance of any schedule 2 quotation.

(2) The amount (either an addition or an omission) of any variation to any schedule 2 quotation.

(3) The deduction of all Provisional Sums and the addition of the amount of the valuation of the works executed by or the disbursements made by the sub-contractor in accordance with the directions of the contractor as to the expenditure of those Provisional Sums.

(4) The deduction of the value of all work described as provisional and included in the sub-contract documents, and the addition of the amount of the valuation of the works executed by or the disbursements made by the sub-contractor in accordance with the directions of the contractor as to the expenditure against all work described as provisional and included in the sub-contract documents.

(5) The deduction of the value of all approximate quantities included in any bill of quantities, and the addition of the amount of the valuation of the works executed by or the disbursements made by the sub-contractor in accordance with the directions of the contractor as to the expenditure against all approximate quantities described in any bill of quantities or in the Contractor's Requirements.

(6) The adjustment for the amount due (whether this be an addition or an omission) of each variation valued in line with the valuation rules, together with the adjustment (whether this be an addition or an omission) of the amount included in the sub-contract documents for any other work which has suffered a substantial change in its conditions due to the variation in question.

(7) The deduction in respect of applicable deductions made under the main contract conditions, and in respect of non-compliant work.

(8) The addition of any payment due to the sub-contractor as a result of payments made or costs incurred by the sub-contractor in relation to fees or charges legally demandable and royalties and patent rights under the main contract conditions.

(9) The amount of any valuation in respect of non-compliant work by others.

(10) The amount ascertained in respect of the sub-contractor's loss and expense (including that relating to the proper suspension of the works by the sub-contractor).

(11) Any adjustment (whether this be an addition or an omission) due to the fluctuation option applicable.

(12) The adjustment (whether this be an addition or an omission) of any other amount, which may, for example, include costs incurred by the contractor that are associated with the non-compliance of directions by the sub-contractor, or associated with insurance taken out on behalf of the sub-contractor.

8.4.2 Final sub-contract sum – re-measurement basis (i.e. sub-contract based upon sub-contract tender sum)

Under the re-measurement basis, the quality and quantity of work included in the sub-contract tender sum is that set out in the bills of quantities and the sub-contractor's design documents, or where there are no bills of quantities, the quality and quantity of work will be that included in the sub-contract documents taken together. If the contractor provides quantities and these are contained in the sub-contract documents, then the quantity of work contained in the sub-contract tender sum shall be in line with those quantities.

When the re-measurement basis applies, the final sub-contract sum is based upon a valuation of a complete re-measurement of the works rather than being based on additions to or omissions from the sub-contract tender sum.

This amount is then adjusted by the following:

(1) The amount stated (either an addition or an omission) by the contractor in his or her acceptance of any schedule 2 quotation.

(2) The amount (either an addition or an omission) of any variation to any schedule 2 quotation.

(3) The addition of any payment due to the sub-contractor as a result of payments made or costs incurred by the sub-contractor in respect of fees or charges legally demandable and royalties and patent rights payments due under the main contract conditions.

(4) The amount of any valuation in respect of non-compliant work by others

(5) The deduction of any applicable deductions under the main contract conditions, and any applicable deductions in respect of non-compliant work.

(6) The amount ascertained in respect of the sub-contractor's loss and expense (including that relating to the proper suspension of the works by the sub-contractor).

(7) Any adjustment (whether this be an addition or an omission) due to the fluctuation option applicable.

(8) The adjustment (whether this be an addition or an omission) of any other amount, which may include for costs incurred by the contractor that are associated with the non-compliance of directions by the sub-contractor or associated with insurance taken out on behalf of the sub-contractor.

8.5 Retention

Retention is a contractual provision permitting the contractor to withhold a sum of money against the value of works executed by the sub-contractor to allow for the rectification of defects.

Post-Practical Completion retention is withheld in respect of latent defects in the sub-contractor's works, that is, defects that are not apparent at practical completion of the sub-contract. Generally, any patent defects (i.e. obvious defects) should be dealt with by way of a reduction in the valuation of the works at the appropriate time.

The specific rules and provisions governing the withholding and release of retention are stated in the sub-contract, along with the retention percentage to be applied to the sub-contractor's interim payments.

8.5.1   Amount of and release of retention

In line with clause 4.15, the retention which may be deducted is to be calculated as follows:

(1) Where the sub-contract works or such works in any Section have not reached practical completion, the retention which the contractor may deduct and retain shall be the percentage stated under item 9 of the sub-contract particulars applied to the part of the gross valuation (appropriate at that time) covered by clauses 4.13.1-4.13.3 inclusive. It should be noted that clauses 4.13.1-4.13.3 only relate to part of the gross valuation, and exclude, amongst other things, loss and expense assessments. Retention is therefore only held on the part of the gross valuation as specified. The retention percentage may be inserted by the parties to the sub-contract under item 9 of the sub-contract particulars; however, if no rate is inserted, the default percentage is 3% (which is the same as the default position under the main contract).

(2) Where the sub-contract works or such works in any Section have reached practical completion, the retention which the contractor may deduct and retain shall be one half of the amount calculated under point (1) above.

(3) On any occasion when the amount of retention applicable through applying points (1) and (2) above falls below the minimum retention amount stated in sub-contract particulars item 9, then no retention will be held at all. A minimum retention amount (above £250) may be inserted by the parties to the sub-contract under item 9 of the sub-contract particulars; however, if no such amount is inserted, the default minimum amount is £250.

(4) Subject to practical completion of the sub-contract works having been achieved, and subject to their being no defects, shrinkages or other faults of the type referred to in clause 2.22.1 apparent at the retention release date, the balance of the retention held will be released in the next interim payment due. If not already previously notified to the sub-contractor, the applicable defects shall be notified to the sub-contractor no later than 14 days after the retention release date. The retention release date is the date specified in the sub-contract particulars under item 9. If no date is inserted, the default position is that the retention release date is to be 6 months after the end of the rectification period from the Main Contract Date for Completion (both as stated in the Main Contract, and as noted under sub-contract particulars item 9). This is a major departure from earlier editions of the JCT sub-contract forms, and, for the first time, the final release of retention to a sub-contractor is entirely independent from the final release of retention to the main contractor.

8.5.2   Retention bond

A retention bond only applies where item 10 of the sub-contract particulars is completed, recording that a retention bond is to apply. The default position is that, unless so marked, a retention bond is not to apply. The retention bond is to be in the form set out under schedule 3 part 2, and the organisation giving the surety on the bond is to approved by the contractor.

The purpose of the retention bond is to give the contractor security that monies can be recovered by way of the retention bond, rather than by holding retention monies against the sub-contractor's account. Despite this, the contractor (at each interim payment) is to prepare a statement which specifies the deduction in respect of retention that would have been made had the retention bond not been in place (as clause 4.16.1), so that if a valid retention bond is not in place, retention monies would be held in the normal manner. If that was to occur, the retention monies held would be released in the next interim payment after a valid retention bond was in place.

The retention bond needs to state the surety's maximum liability amount under the bond, and needs to state the expiry date of the bond; and both the amount and the date needs to be inserted in the sub-contract particulars under item 10.

If at any time, the retention monies that would have been held if a retention bond had not been in place exceeds the maximum aggregate sum of the retention bond, then either the sub-contractor will arrange for the sum in retention bond to be increased by the surety to a revised surety level to suit or the residual amount of retention monies in excess of the maximum aggregate sum of the retention bond will be held in the normal way against the sub-contractor's account.

The purpose of the retention bond is to allow the contractor to recover from the surety:

(1)   the costs actually incurred by the contractor by reason of the failure of the sub-contractor to comply with the directions of the contractor under the sub-contract (as clause 4.3.1 of the retention bond);

(2)   any expenses or any direct loss and/or damage caused to the contractor as a result of the termination of the sub-contractor's employment by the contractor (as clause 4.3.2 of the retention bond);

(3)   any costs, other than the amounts referred to in points (1) and (2) above, which the contractor has actually incurred and which, under the sub-contract, he or she is entitled to deduct from monies otherwise due or to become due to the sub-contractor (as clause 4.3.3 of the retention bond).

The second paragraph to item 3 of the notes to the retention bond states that any demand under clause 4 of the retention bond must not exceed the costs actually incurred by the contractor; therefore (and accordingly), the contractor cannot add profit to any costs incurred when making a claim under the retention bond.

Before making a claim under the retention bond, the contractor is required to give written notice to the sub-contractor of his or her liability for the amount demanded, and is to request the sub-contractor to discharge his or her liability. At the time when such a notice is sent to the sub-contractor, a copy is to be sent to the surety of the retention bond. A claim under the retention bond cannot be made until at least 14 days after the said written notice has been provided to the sub-contractor, and can only be made if the sub-contractor has not complied with the requirements of the written notice.

The last paragraph of clause 4 of the retention bond states:

> 'Such demand as above shall, for the purposes of this Bond but not further or otherwise, be conclusive evidence (and admissible as such) that the amount demanded is properly due and payable to the Contractor by the Sub-contractor'.

In other words, a demand made under clauses 4.3.1–4.3.3 of the retention bond will be taken as being conclusive evidence to the surety that the amount demanded is properly due and payable (therefore, the retention bond is, in effect, an 'on-demand' bond), but any such demand will not prevent the sub-contractor from pursuing his or her claim under the terms of the sub-contract separately.

Finally, clause 4.16.5 of SBCSub/D/C notes that where the contractor has required the sub-contractor to provide a performance bond, then, in respect of any default to which that performance bond refers, which is also a matter for which the contractor could make, under the terms of the retention bond, the contractor shall first have recourse to the retention bond.

8.6 Payment due dates and final dates for payment for interim payments

8.6.1 Payment due date for the first interim payment

Clause 4.9.1.1 notes that the first interim payment is due on either:

- the date specified in the sub-contract particulars in item 7; or
- if no such date is specified, and unless otherwise agreed in writing, the date would be the date 1 month after the date of commencement of the sub-contract works on site.

Where clause 4.9.2 applies (as stated under sub-contract particulars item 7, but where if no entry is made, clause 4.9.2 is deemed not to apply), then the payment due date for the first interim payment shall be as set out above for clause 4.9.1.1, *or if later* shall be the date of receipt by the contractor of an application for payment by the sub-contractor which states the sum the sub-contractor considers will become or to have been due to him or her in accordance with clause 4.9.4 at the due date and which states the basis on which that sum has been calculated. Therefore, under clause 4.9.2, a payment does not become due until after a sub-contractor makes a payment application.

8.6.2 Payment due date for subsequent interim payments

Where clause 4.9.2 does not apply, clause 4.9.1.2 notes that the payment due dates for subsequent interim payments in the period up to and including the month following practical completion of the main contract works are on the same date in each month following the date of the payment due date for the first interim payment. If the first due date is a date that does not recur in a subsequent month, the due date for a subsequent month shall be the last day of that month (as clause 4.9.1). After practical completion of the main contract works (and up to the last payment due date), the payment due dates shall be as for all other interim payments other than that the interval between the interim payments shall increase from 1 month to 2 months.

Again, where clause 4.9.2 applies (as stated under sub-contract particulars item 7, but where if no entry is made, clause 4.9.2 is deemed not to apply), the payment due date for the subsequent interim payments shall be as set out above for clause 4.9.1.2, *or if later* shall be the date of receipt by the contractor of an application for payment by the sub-contractor which states the sum the sub-contractor considers will become or to have been due to him or her in accordance with clause 4.9.4 at the due date and which states the basis on which that sum has been calculated.

8.6.3   The final date for payment of an interim payment

The final date for payment of an interim payment is 21 days after the date on which those payments become due (as clause 4.10.1).

8.7   Payment Notices and Pay Less Notices for interim payments

In line with the Local Democracy, Economic Development and Construction Act 2009, a contractor is required to pay the notified sum stated on either a Payment Notice or a Pay Less Notice. In addition, the courts have found in a recent court case[3] that a payer (an employer in that particular case, but the contractor in respect of the JCT sub-contracts under consideration in this book) must serve the required Pay Less Notice, otherwise he or she will have to pay the sum the payee (a contractor in that case, but the sub-contractor in respect of the JCT sub-contracts under consideration in this book) applied for. Therefore, now, more than ever, the timing and content of Payment Notices, Default Payment Notices and Pay Less Notices is critical in respect of payments due to be made to sub-contractors.

8.7.1   Contractor's Payment Notice

Clause 4.10.2 says that the contractor is to give a Payment Notice to the sub-contractor not later than 5 days after the due date specifying the sum that he or she considers to be due or to have been due to the sub-contractor at the payment due date. This notice is to be given even if it is considered that the amount

---

[3]*ISG Construction Ltd* v. *Seevic College* [2014] EWHC 4007.

due is zero (as clause 4.10.9). Crucially, it is noted that, subject to any Pay Less Notice given by the contractor under clause 4.10.5, the sum to be paid by the contractor shall be the sum specified in the Payment Notice.

8.7.2   Contractor's Pay Less Notice

Clause 4.10.5 says that if the contractor intends to pay less than the sum stated as due from him or her in his or her Payment Notice (pursuant to clause 4.10.2), he or she is, not later than 5 days before the final date for payment, to give a Pay Less Notice which (as required by clause 4.10.8.1) is to specify both the sum that he or she considers to be due at the date the Pay Less Notice is given and the basis on which that sum has been calculated. This notice is to be given even if it is considered that the amount due is zero (as clause 4.10.9).

8.7.3   Sub-contractor's Default Payment Notice

A sub-contractor must or may make a payment application in the following cases:

- If clause 4.9.2 applies, a sub-contractor must make a payment application, otherwise a payment due date cannot be set.
- If clause 4.9.2 does not apply, a sub-contractor may make a payment application at any time not later than 7 days prior to the relevant payment due date (in line with clause 4.9.3.1);
- In line with clause 4.9.3.2, a sub-contractor may make a payment application at any time after the 5-day period referred to in clause 4.10.2 for the contractor to make his or her Payment Notice if the contractor fails to give such a Payment Notice and the sub-contractor has not in relation to that payment already made a payment application under clause 4.9.3.1.

If the contractor fails to give a Payment Notice in accordance with clause 4.10.2, then (other than where the contractor issues a Pay Less Notice pursuant to clause 4.10.5) the sum to be paid by the contractor to the sub-contractor is the sum specified in the sub-contractor's payment application (i.e. the sub-contractor's Default Payment Notice), which is made by the sub-contractor in one of the ways noted above.

If a payment application is made by a sub-contractor in accordance with clause 4.9.3.2, the final date for payment for the payment in question is postponed by the same number of days as the number of days after the expiry of the 5-day period referred to in clause 4.10.2 (when the contractor was required to make his or her own Payment Notice) that the payment application is made. Therefore, if the sub-contractor's payment application was made 6 days after the date when the contractor's Payment Notice should have been issued pursuant to clause 4.10.2, then the final date for payment would be postponed by 6 days also. In such a situation, the contractor's Pay Less Notice relates to the revised final date for payment rather than the original final date for payment.

8.8 The payment due date and the final date for payment for the final payment

8.8.1 The payment due date for the final payment

Clause 4.12.1 says that the payment due date for the final payment is whichever of the following that occurs last:

- the retention release date
- where applicable, any later due date for retention release pursuant to clause 4.15.2.3
- the date of issue of the contractor's statement under clause 4.6.2 or, in default, the last date for issue of that statement (i.e. the date of issue of the Final Certificate under the main contract).

8.8.2 The final date for payment of the final payment

Subject only to clause 4.12.6, the final date for payment for the final payment is 28 days after its due date.

8.9 Payment Notices and Pay Less Notices in respect of the final payment

In line with the Local Democracy, Economic Development and Construction Act 2009, a contractor is required to pay the notified sum stated on either a Payment Notice or a Pay Less Notice. In addition, the courts have found in a recent court case[4] that a payer (an employer in that particular case, but the contractor in respect of the JCT sub-contracts under consideration in this book) must serve the required Pay Less Notice, otherwise he or she will have to pay the sum the payee (a contractor in that case, but the sub-contractor in respect of the JCT sub-contracts under consideration in this book) applied for. Therefore, now, more than ever, the timing and content of Payment Notices, Default Payment Notices and Pay Less Notices is critical in respect of payments due to be made to sub-contractors.

Also, reference must be made to the commentary in respect of the conclusive effect of the Final Payment Notice.

8.9.1 Contractor's Final Payment Notice

Clause 4.12.2 says that the contractor is to give the Final Payment Notice to the sub-contractor not later than 5 days after the due date which specifies the final sub-contract sum calculated in accordance with whichever of clause 4.3 or clause 4.4 applies less the total amount previously due as interim payments. The

---

[4]*ISG Construction Ltd* v. *Seevic College* [2014] EWHC 4007.

balance remaining shall be shown as a balance due from the contractor to the sub-contractor, or as a balance due from the sub-contractor to the contractor, as applicable.

8.9.2   Contractor's or sub-contractor's Pay Less Notice in respect of the contractor's Final Payment Notice

Clause 4.12.4 says that if the contractor (or the sub-contractor, as applicable) intends to pay less than the sum stated as due from him or her in the contractor's Final Payment Notice (pursuant to clause 4.12.2), he or she is, not later than 5 days before the final date for payment, to give a Pay Less Notice which (as required by clause 4.10.8.1) is to specify both the sum that he or she considers to be due at the date the notice is given and the basis on which that sum has been calculated. This notice is to be given even if it is considered that the amount due is zero (as clause 4.10.9). Clause 4.12.5 notes that where a Pay Less Notice is given, the payment to be made on or before the final date for payment shall not be less than the amount stated as due in the Pay Less Notice.

8.9.3   Sub-contractor's default Final Payment Notice

If the contractor fails to give a Final Payment Notice in accordance with clause 4.12.2, then the sub-contractor may (pursuant to clause 4.12.6) give a default Final Payment Notice. Other than where the contractor issues a Pay Less Notice in accordance with clauses 4.12.4 and 4.10.8, the sum to be paid is as stated in the sub-contractor's default Final Payment Notice.

If a default Final Payment Notice is made by a sub-contractor in accordance with clause 4.12.6.1, the final date for payment for the payment in question is postponed by the same number of days as the number of days after the expiry of the 5-day period referred to in clause 4.12.2 (when the contractor was required to make his or her own Final Payment Notice) that the sub-contractor's Final Payment Notice is made. Therefore, if the sub-contractor's default Final Payment Notice was made 6 days after the date when the contractor's Final Payment Notice should have been issued pursuant to clause 4.12.2, then the final date for payment would be postponed by 6 days also. In such a situation, the contractor's Pay Less Notice relates to the revised final date for payment rather than the original final date for payment.

8.9.4   Contractor's Pay Less Notice in respect of the sub-contractor's default Final Payment Notice

Clause 4.12.6.3 says that if the contractor intends to pay less than the sum stated as due from him or her in the sub-contractor's default Final Payment Notice (pursuant to clause 4.12.6.1), he or she is, not later than 5 days before the final date for payment, to give a Pay Less Notice which (as required by clause 4.10.8.1) is to specify both the sum that he or she considers to be due at the date the notice is given and the basis on which that sum has been calculated.

This notice is to be given even if it is considered that the amount due is zero (as clause 4.10.9). Clause 4.12.5 notes that where a Pay Less Notice is given, the payment to be made on or before the final date for payment shall not be less than the amount stated as due in the Pay Less Notice.

8.10   VAT (Value Added Tax)

Clause 4.7.1 of the sub-contract conditions makes it clear that the sub-contract sum, the sub-contract tender sum and the final sub-contract sum are exclusive of value added tax (VAT), and adds that in relation to any payment made to the sub-contractor under the sub-contract, the contractor shall pay the amount of VAT properly chargeable in respect of that payment.

If, after the sub-contract base date, the supply of any goods and services to the contractor by the sub-contractor becomes exempt from VAT and the VAT therefore cannot be recovered by the sub-contractor, the contractor is to pay the sub-contractor an amount equal to the amount of the VAT input tax on the supply to the sub-contractor of the goods and services which contribute to the appropriate part of the sub-contract works but which, because of the exemption, the sub-contractor cannot recover.

This rule only applies where the supply of any goods and services to the contractor by the sub-contractor becomes exempt from VAT *after* the sub-contract base date. The sub-contract base date is given in the sub-contract particulars.

8.11   Construction Industry Scheme (CIS)

Clause 4.8 notes that the obligation of the contractor to make any payment under the sub-contract is subject to the provisions of the Construction Industry Scheme (CIS). Under the CIS, contractors deduct money from a sub-contractor's payments and pass it to HM Revenue and Customs (HMRC). The deductions count as advance payments towards the sub-contractor's tax and National Insurance. Contractors must register for the scheme. Sub-contractors do not have to register, but deductions are taken from their payments at a higher rate if they are not registered.

8.12   Interest

Interest means a financial charge that is made against another party in respect of outstanding monies owed. Prior to the Late Payment of Commercial Debts (Interest) Act 1998, a party had no automatic right or implied right to interest unless there was an express term in the contract which dealt with the question of interest. However, there is now an implication that in respect of commercial contracts and sub-contracts debts carries a right to simple interest and a right to financial compensation.

Under the SBCSub/D/C, interest is expressly dealt with under clauses 4.10.6 (for interim payments) and 4.12.7 (for final payments).

In respect of both clause 4.10.6 and clause 4.12.7, the sub-contract states that if the party that is due to make payment (i.e. the payer in this context) fails properly to pay the amount due, or any part of it, to the other party by the final date of its payment, the payer shall pay to the other party, in addition to the amount not properly paid, simple (not compound) interest thereon at the interest rate for the period until such payment is made. In respect of clause 4.10.6, the interest due shall be treated as a debt due to the sub-contractor from the contractor, and may be pursued by the sub-contractor in the same way as any other debt due under the sub-contract. The acceptance of any payment of interest by the sub-contractor shall not in any circumstances be construed as a waiver by the sub-contractor of his or her right to proper payment of the principal amounts due from the contractor, and (in respect of clause 4.10.6) it will not be construed as a waiver to suspend performance of his or her obligations, or to terminate his or her employment under the sub-contract.

The interest rate that applies (as defined under clause 1.1) is 'a rate 5% per annum above the official dealing rate of the Bank of England correct at the date that a payment due under this sub-contract becomes overdue'.

From the above, it appears that the interest rate applicable to a particular overdue payment is fixed, (i.e. at 5% per annum above the official dealing rate of the Bank of England) at the date that the payment became due, and does not, as one might expect, change during the period of time that the debt remains outstanding to suit any fluctuations (whether these be up or down) to the official dealing rate of the Bank of England.

8.13   Sub-contractor's right of suspension

The sub-contractor's right of suspension is dealt with under clause 4.11.

If the contractor fails to pay the sub-contractor the sum payable in respect of an interim payment, together with any VAT properly chargeable in respect of that payment, by the final date for payment, then, if that non-payment continues for 7 days after the sub-contractor has given notice of his or her intention to suspend performance of his or her obligations under this sub-contract and the ground or grounds on which it is intended to suspend performance, the sub-contractor may suspend performance of any or all of his or her obligations until payment is made in full. There is no guidance as to when 'payment is made in full' occurs; however, in most cases, this should be obvious. In terms of cheques, however, it is a well-settled principle of law that a cheque is to be treated as cash, and thus is not susceptible to set off.[5] It is, therefore, considered that the courts may decide that 'payment in full' is made when the sub-contractor receives a cheque rather than (the later date) when the cheque had been cleared through the sub-contractor's account.

It is important to note that the trigger for validly suspending performance runs from the date of the notice of suspension, not from the date when payment should have been made. The notice of suspension may often be issued some time after the final date for payment, not least due to promises of 'the cheque is in the post'.

The sub-contractor can only suspend in respect of sums that are payable, not in respect of sums that a sub-contractor has claimed, for which may not in fact be payable. In this regard, it must be noted that a wrongful suspension by the sub-contractor may amount to a repudiation of the sub-contract by the sub-contractor, which, if accepted by the contractor, may lead to the sub-contractor suffering all of the contractor's damages arising from that repudiation. It may also constitute a ground for termination of the sub-contract (e.g. see clause 7.4.1.1, i.e. 'without reasonable cause wholly or substantially suspends the carrying out of the Sub-contract Works …').

The sub-contractor's right to suspend performance does not affect any other rights and remedies the sub-contractor holds under the sub-contract. By way of example, the sub-contractor would be entitled to give notice of termination under the contract as well as suspend performance. The sub-contractor would also be entitled to claim interest on late payment.

If the sub-contractor does (correctly) suspend performance, he or she is entitled to an extension of time to the period for completion pursuant to the Relevant Sub-contract Event covered by clause 2.19.6. The extension of time period will be the period when the suspension starts (which cannot occur until after the expiry of the 7-day notice period) to the time when payment in full occurs. Unfortunately, there is no clear guidance as to whether or not any required re-mobilisation period by the sub-contractor would be recoverable by the sub-contractor upon his or her return to site as part of its extension of time entitlement (although it is felt that some allowance for this should be made).

Clause 4.11.2 notes that where the sub-contractor exercises his or her right of suspension (as set out above), he or she is entitled to a reasonable amount in respect of costs and expenses reasonably incurred by him or her as a result of the exercise of the right. The sub-contractor is to make an application for the relevant costs and expenses, and either with that application or following a request by the contractor, is to provide such details of the costs and expenses claimed as are reasonably necessary to enable the sub-contractor's entitlement to be ascertained.

8.14 Fluctuations

Under clause 4.17, three fluctuation provisions are referred to:

■ Fluctuations option A: contribution, levy and tax fluctuations
■ Fluctuations option B: labour and materials cost, and tax fluctuations
■ Fluctuations option C: formula adjustment

The fluctuations option which is applicable to any particular sub-contract is indicated under item 11 of the sub-contract particulars. Footnote 22 to item 11 of the sub-contract particulars states that all but one fluctuations option is to be deleted. Therefore, the intention is that one fluctuations option (and only one fluctuations option) will apply to every sub-contract. Footnote 21 to item 10 of the sub-contract particulars states that where fluctuations option A or B applies in the main contract, it is a requirement of that contract that the same option applies to any sub-contract.

Clause 4.18 notes that, irrespective of which fluctuations option applies to the sub-contract works, generally, none of the fluctuations options will apply in respect of work for which a schedule 2 quotation has been accepted by the contractor or in respect of a variation to an accepted schedule 2 quotation.

Details of the three fluctuations options are set out in schedule 4 to the SBCSub/D/C.

8.15   Equivalent sub-contract provisions

In the table below, the equivalent sub-contract provisions to that within the SBCSub/D/A or SBCSub/D/C in the text above are listed for the SBCSub/A or SBCSub/C, DBSub/A or DBSub/C, ICSub/A or ICSub/C and ICSub/D/A or ICSub/D/C.

SBCSub/D/A or SBCSub/D/C	SBCSub/A or SBCSub/C	DBSub/A or DBSub/C	ICSub/A or ICSub/C	ICSub/D/A or ICSub/D/C
Article 3A	Article 3A	Article 3A	Article 3A	Article 3A
Article 3B	Article 3B	Article 3B	Article 3B	Article 3B
Schedule of Particulars Item 7	Schedule of Particulars Item 7	Schedule of Particulars Item 7	Schedule of Particulars Item 7	Schedule of Particulars Item 7
Sub-contract Particulars Item 8	Sub-contract Particulars Item 8	Sub-contract Particulars Item 8	Sub-contract Particulars Item 9	Sub-contract Particulars Item 9
Schedule of Particulars Item 9	Schedule of Particulars Item 9	Schedule of Particulars Item 9	Schedule of Particulars Item 8	Schedule of Particulars Item 8
Sub-contract Particulars Item 10	Sub-contract Particulars Item 10	Sub-contract Particulars Item 10	Not applicable to this Sub-contract as this Sub-contract Particular relates to a Retention Bond, which is not provided for under this Sub-contract	Not applicable to this Sub-contract as this Sub-contract Particular relates to a Retention Bond, which is not provided for under this Sub-contract
Schedule 2 Quotation	Schedule 2 Quotation	Schedule 2 Quotation	Not applicable to this Sub-contract as this Sub-contract Particular relates to a Schedule 2 Quotation, which is not provided for under this Sub-contract	Not applicable to this Sub-contract as this Sub-contract Particular relates to a Schedule 2 Quotation, which is not provided for under this Sub-contract
Clause 1.1	Clause 1.1	Clause 1.1	Clause 1.1	Clause 1.1

(Continued)

SBCSub/D/A or SBCSub/D/C	SBCSub/A or SBCSub/C	DBSub/A or DBSub/C	ICSub/A or ICSub/C	ICSub/D/A or ICSub/D/C
Clause 2.15.1	Clause 2.15.1	Clause 2.15.1	Clause 2.11.1	Clause 2.11.1
Clause 2.15.2	Clause 2.15.2	Clause 2.15.2	Clause 2.11.2	Clause 2.11.2
Clause 2.15.3	Clause 2.15.3	Clause 2.15.3	Clause 2.11.3	Clause 2.11.3
Clause 2.19.6	Clause 2.19.6	Clause 2.19.6	Clause 2.13.6	Clause 2.13.6
Clause 2.22.1	Clause 2.22.1	Clause 2.22.1	Clause 2.16.1	Clause 2.16.1
Clause 4.3	Clause 4.3	Clause 4.3	Clause 4.3	Clause 4.3
Clause 4.3.1	Clause 4.3.1	Clause 4.3.1	Clause 4.3.1 (Variation and Acceleration Quotations excluded)	Clause 4.3.1 (Variation and Acceleration Quotations excluded)
Clause 4.3.2	Clause 4.3.2	Clause 4.3.2	Clause 4.3.2	Clause 4.3.2
Clause 4.3.3	Clause 4.3.3	Clause 4.3.3	Clause 4.3.3	Clause 4.3.3
Clause 4.4	Clause 4.4	Clause 4.4	Clause 4.4 (Variation and Acceleration Quotations excluded)	Clause 4.4 (Variation and Acceleration Quotations excluded)
Clause 4.6.1	Clause 4.6.1	Clause 4.6.1	Clause 4.6.1	Clause 4.6.1
Clause 4.6.2	Clause 4.6.2	Clause 4.6.2 (Clause 4.6.3 added)	Clause 4.6.2	Clause 4.6.2
Clause 4.7.1	Clause 4.7.1	Clause 4.7.1	Clause 4.7.1	Clause 4.7.1
Clause 4.8	Clause 4.8	Clause 4.8	Clause 4.8	Clause 4.8
Clause 4.9.1	Clause 4.9.1	Clause 4.9.1	Clause 4.9.1 (clause 4.9.1.3 added)	Clause 4.9.1 (clause 4.9.1.3 added)
Clause 4.9.1.1	Clause 4.9.1.1	Clause 4.9.1.1	Clause 4.9.1.1	Clause 4.9.1.1
Clause 4.9.1.2	Clause 4.9.1.2	Clause 4.9.1.2	Clause 4.9.1.2 (to be read with clause 4.9.1.3)	Clause 4.9.1.2 (to be read with clause 4.9.1.3)
Clause 4.9.2	Clause 4.9.2	Clause 4.9.2	Not applicable to this Sub-contract as this clause relates to a Payee-led (i.e. a Sub-contractor-led) payment regime, which is not provided for under this Sub-contract	Not applicable to this Sub-contract as this clause relates to a Payee-led (i.e. a Sub-contractor-led) payment regime, which is not provided for under this Sub-contract
Clause 4.9.3.1	Clause 4.9.3.1	Clause 4.9.3.1	Not applicable to this Sub-contract as this clause relates to a Payee-led (i.e. a Sub-contractor-led) payment regime, which is not provided for under this Sub-contract	Not applicable to this Sub-contract as this clause relates to a Payee-led (i.e. a Sub-contractor-led) payment regime, which is not provided for under this Sub-contract

SBCSub/D/A or SBCSub/D/C	SBCSub/A or SBCSub/C	DBSub/A or DBSub/C	ICSub/A or ICSub/C	ICSub/D/A or ICSub/D/C
Clause 4.9.3.2	Clause 4.9.3.2	Clause 4.9.3.2	Not applicable to this Sub-contract as this clause relates to a Payee-led (i.e. a Sub-contractor-led) payment regime, which is not provided for under this Sub-contract	Not applicable to this Sub-contract as this clause relates to a Payee-led (i.e. a Sub-contractor–led) payment regime, which is not provided for under this Sub-contract
Clause 4.9.4	Clause 4.9.4	Clause 4.9.4	Clauses 4.9.3 and 4.9.4 combined	Clauses 4.9.3 and 4.9.4 combined
Clause 4.10.1	Clause 4.10.1	Clause 4.10.1	Clause 4.12.1	Clause 4.12.1
Clause 4.10.2	Clause 4.10.2	Clause 4.10.2	Clause 4.12.2	Clause 4.12.2
Clause 4.10.5	Clause 4.10.5	Clause 4.10.5	Clause 4.12.5	Clause 4.12.5
Clause 4.10.6	Clause 4.10.6	Clause 4.10.6	Clause 4.12.6	Clause 4.12.6
Clause 4.10.8	Clause 4.10.8	Clause 4.10.8	Clause 4.12.8	Clause 4.12.8
Clause 4.10.8.1	Clause 4.10.8.1	Clause 4.10.8.1	Clause 4.12.8.1	Clause 4.12.8.1
Clause 4.10.9	Clause 4.10.9	Clause 4.10.9	Clause 4.12.9	Clause 4.12.9
Clause 4.11	Clause 4.11	Clause 4.11	Clause 4.13	Clause 4.13
Clause 4.11.2	Clause 4.11.2	Clause 4.11.2	Clause 4.13.2	Clause 4.13.2
Clause 4.12.1	Clause 4.12.1	Clause 4.12.1	Clause 4.14.1 (but worded differently)	Clause 4.14.1 (but worded differently)
Clause 4.12.2	Clause 4.12.2	Clause 4.12.2	Clause 4.14.2	Clause 4.14.2
Clause 4.12.4	Clause 4.12.4	Clause 4.12.4	Clause 4.14.4	Clause 4.14.4
Clause 4.12.5	Clause 4.12.5	Clause 4.12.5	Clause 4.14.5	Clause 4.14.5
Clause 4.12.6	Clause 4.12.6	Clause 4.12.6	Clause 4.14.6	Clause 4.14.6
Clause 4.12.6.1	Clause 4.12.6.1	Clause 4.12.6.1	Clause 4.14.6.1	Clause 4.14.6.1
Clause 4.12.6.3	Clause 4.12.6.3	Clause 4.12.6.3	Clause 4.14.6.3	Clause 4.14.6.3
Clause 4.12.7	Clause 4.12.7	Clause 4.12.7	Clause 4.14.7	Clause 4.14.7
Clause 4.13	Clause 4.13	Clause 4.13	Clause 4.10 (but worded differently)	Clause 4.10 (but worded differently)
Clause 4.13.1	Clause 4.13.1	Clause 4.13.1	Clause 4.10.1 to 4.10.3 (but worded differently)	Clause 4.10.1 to 4.10.3 (but worded differently)
Clause 4.13.2	Clause 4.13.2	Clause 4.13.2	Clause 4.10.1 to 4.10.3 (but worded differently)	Clause 4.10.1 to 4.10.3 (but worded differently)
Clause 4.13.3	Clause 4.13.3	Clause 4.13.3	Clause 4.10.1 to 4.10.3 (but worded differently)	Clause 4.10.1 to 4.10.3 (but worded differently)

(Continued)

SBCSub/D/A or SBCSub/D/C	SBCSub/A or SBCSub/C	DBSub/A or DBSub/C	ICSub/A or ICSub/C	ICSub/D/A or ICSub/D/C
Clause 4.13.9	Clause 4.13.9	Clause 4.13.9	Clause 4.10.1 to 4.10.3 (but worded differently)	Clause 4.10.1 to 4.10.3 (but worded differently)
Clause 4.15	Clause 4.15	Clause 4.15	Dealt with under clause 4.10	Dealt with under clause 4.10
Clause 4.15.2.3	Clause 4.15.2.3	Clause 4.15.2.3	Dealt with under clause 4.10	Dealt with under clause 4.10
Clause 4.16.1	Clause 4.16.1	Clause 4.16.1	Not applicable to this Sub-contract as this clause relates to a Retention Bond, which is not provided for under this Sub-contract	Not applicable to this Sub-contract as this clause relates to a Retention Bond, which is not provided for under this Sub-contract
Clause 4.16.5	Clause 4.16.5	Clause 4.16.5	Not applicable to this Sub-contract as this clause relates to a Performance Bond, which is not provided for under this Sub-contract	Not applicable to this Sub-contract as this clause relates to a Performance Bond, which is not provided for under this Sub-contract
Clause 4.17	Clause 4.17	Clause 4.17	Clause 4.15	Clause 4.15
Clause 4.18	Clause 4.18	Clause 4.18	Not applicable to this Sub-contract as this clause relates to the fluctuations in respect of a schedule 2 Quotation, and a schedule 2 Quotation is not provided for under this Sub-contract	Not applicable to this Sub-contract as this clause relates to the fluctuations in respect of a schedule 2 Quotation, and a schedule 2 Quotation is not provided for under this Sub-contract
Clause 7.4.1.1	Clause 7.4.1.1	Clause 7.4.1.1	Clause 7.4.1.1	Clause 7.4.1.1
Schedule 3 Part 1	Schedule 3 Part 1	Schedule 3 Part 1	Schedule 1	Schedule 1
Schedule 3 Part 2	Schedule 3 Part 2	Schedule 3 Part 2	Not applicable to this Sub-contract as this Part of this Schedule relates to a Retention Bond, which is not provided for under this Sub-contract	Not applicable to this Sub-contract as this Part of this Schedule relates to a Retention Bond, which is not provided for under this Sub-contract
Schedule 4	Schedule 4	Schedule 4	Schedule 2	Schedule 2

9 Loss and Expense

9.1 Introduction

In the context of this book loss and expense claims are financial claims submitted by the sub-contractor associated with the prolongation and/or disruption of the works, where the recovery of such additional costs cannot be made under any other provision of the sub-contract.

When considering loss and expense, the first thing that must be noted is that claims in the construction industry generally fall under two differing legal heads. That is:

(1) claims for breach of contract or for other events for which specific provision is made within the conditions of the contract or sub-contract (normally referred to as loss and expense claims);
(2) claims for breach of contract for which no specific provision has been included in the conditions of contract or sub-contract (normally referred to as common law damages claims).

Loss and expense claims have certain advantages over common law damages claims, and some of these advantages are:

- There is a clear definition of responsibilities and procedures.
- There is a clarification of the events and the circumstances for claims to be made.
- There is a right to interim payments and/or decisions in respect of loss and expense.
- There is the avoidance of the uncertainties involved in claiming at common law.
- Loss and expense claims create a right under the contract to a debt, rather than a claim for damages.

Normally, the inability to recover monies under the express terms of the sub-contract for reasons such as not conforming with the procedures required does not preclude a claim at common law on the same matters, provided there is the entitlement at common law to claim damages, and provided that the sub-contractor's common law rights have not been specifically excluded by the sub-contract.

The JCT 2011 Building Sub-contracts, First Edition. Peter Barnes and Matthew Davies.
© 2016 John Wiley & Sons, Ltd. Published 2016 by John Wiley & Sons, Ltd.

If a matter stated in the sub-contract as giving rise to an entitlement to loss and expense is a breach of contract, then the sub-contractor will normally have a right to common law damages. An example of this would be the late issue of information by the contractor.

However, if the matter relied on does not amount to a breach of contract (i.e. issuing variations), then common law damages may not be applicable.

On the other hand, the preservation of the sub-contractor's common law rights may entitle the sub-contractor to make common law damages claims in respect of matters for which there are no express provisions within the loss and expense clauses (e.g. the breach of any applicable implied terms by the contractor).

The authority on the application of damages arising from the contract both in addition to and, where appropriate, instead of the express provisions of the contract for loss and expense is clearly stated in the case of *Stanley Hugh Leach* v. *Merton*.[1]

It is important to note that should the sub-contractor need to pursue his or her claim at common law because he or she has not followed the procedures under the sub-contract, then it is possible that he or she may be penalised by the court if his or her non-conformity with the sub-contract procedure did not allow the contractor to mitigate his or her loss.

What this means in practice is that if the contractor had been made aware of the pending claim at an earlier stage (in line with the express provisions of the sub-contract), he or she may have been able to have taken steps to reduce or even eliminate the effects of the breach. If the contractor was not given the opportunity to mitigate his or her loss in this way, the sub-contractor may have any future award in damages reduced by a court because of this failing.

A claim for loss and/or expense is the means of putting a sub-contractor back into the position in which he or she would have been but for the delay or disruption; it is not a means of turning a loss into a profit (unless, of course, the loss is because of the non-payment of loss and/or expense). The settlement in effect amounts to common law damages.

Against this background, in this chapter, the following matters will be dealt with, using the sub-contract clause references, etc. contained in the SBCSub/D/A or SBCSub/D/C (as appropriate). Whilst it is clearly beyond the scope of this book to review every nuance of the other sub-contract forms under consideration, the equivalent provisions (where applicable) within the SBCSub/A or SBCSub/C, the DBSub/A or DBSub/C, the ICSub/A or ICSub/C, and the ICSub/D/A or ICSub/D/C are given at the table at the end of this chapter. It must be emphasised that before considering a particularly issue, the actual terms of the appropriate edition of the relevant sub-contract should be reviewed by the reader (and/or legal advice should be sought as appropriate) before proceeding with any action/inaction in respect of the sub-contract in question.

---

[1]*Stanley Hugh Leach* v. *London Borough of Merton*(1985) 32 BLR 51.

In this chapter the following matters are dealt with:

■ Can common law damages claims be excluded by the contract?
■ Are claims for extensions of time and loss and/or expense linked?
■ What are the grounds/requirements for loss and expense?
■ What are Relevant Sub-contract Matters?
■ Can the contractor recover direct loss and/or expense from the sub-contractor?
■ In pursuing a loss and expense claim, what does a sub-contractor need to prove?
■ What needs to be proved in a loss and expense claim document?
■ What is a global claim?
■ Common heads of a loss and expense claim
■ Prolongation costs
■ Disruption claims
■ Winter working
■ Head office overheads and profit
■ Loss of profit
■ Increased costs
■ Cost of claim preparation
■ Interest and finance charges
■ Acceleration
■ Common law damages

9.2 Can common law damages claims be excluded by the contract?

In some circumstances (albeit relatively uncommon), the parties may seek to exclude (via a contract term) the parties' common law rights, that is, so that only the terms of the contract itself governs the relationship between the parties.

However, it must be remembered that the courts view this course of action with some caution, and consequently, there must be a very clear worded contract term in the contract excluding the parties' common law rights for it to be effective. For the avoidance of any doubt, in respect of the above, clause 4.22 preserves the common law damages rights and the remedies of the contractor and the sub-contractor.

9.3 Are claims for extensions of time and loss and/or expense linked?

People in the construction industry often make the mistake of (contractually) linking time and money, in that they assume that financial claim recovery can *only* be secured if an extension of time has been granted and, conversely, that it will *certainly* be due if an extension has been given.

In reality, neither of these propositions is entirely accurate.

Extension of time clauses and those for additional financial recovery (i.e. for loss and expense) are separate matters in the JCT sub-contracts considered in this book, and indeed in all other JCT standard forms.

Additional monies may be recovered even if prolongation is not present, or alternatively, the sub-contractor may receive a comprehensive extension of time and still not be entitled to additional financial recovery.

Therefore, it is advisable that delay and money notices are kept separate.

9.4   What are the grounds/requirements for loss and expense?

Clauses 4.19–4.21 provide for the payment of direct loss and/or expense.

Clause 4.19 provides that if the regular progress of the sub-contract works is materially affected or is likely to be materially affected by any of the Relevant Sub-contract Matters set out in clause 4.20, the sub-contractor may make written application to the contractor for the recovery of the resultant loss and expense incurred.

A sub-contractor is not permitted to recover his or her additional costs twice. Accordingly, clause 4.19 states that any such loss and expense must be that which the sub-contractor would not be reimbursed by a payment under any other provision in the sub-contract (e.g. loss and/or expense would not be recoverable under these provisions where it is being paid elsewhere as part of an accepted schedule 2 quotation).

Assuming that the relevant criterion for claiming loss and/or expense applies, then the contractual provisions are triggered by a written application being made by the sub-contractor. Accordingly, if the sub-contractor makes a written application for loss and expense, then the amount of the payable loss and expense agreed between the contractor and the sub-contractor shall be taken into account in the calculation of the final sub-contract sum or shall be recoverable by the sub-contractor from the contractor as a debt, provided that:

- the sub-contractor's application is made as soon as it has become, or should reasonably have become, apparent to him or her that the regular progress has been or is likely to be affected;
- the sub-contractor submits (upon request from the contractor) such information as is reasonably necessary to show that the regular progress has been or is likely to be affected;
- the sub-contractor submits details of the loss and expense as reasonably requested by the contractor.

9.5   What are Relevant Sub-contract Matters?

The JCT sub-contract provisions regarding loss and expense only provide an entitlement where it can be shown that a Relevant Sub-contract Matter has affected the regular progress of the sub-contract works.

The Relevant Sub-contract Matters are set out in clause 4.20. The Relevant Sub-contract Matters, by and large, replicate some (but not all) of the Relevant Sub-contract Events detailed under clause 2.19 of the sub-contract.

These Relevant Sub-contract Matters are:

- Clause 4.20.1: 'Variations (excluding any for which a Variation Quotation has been accepted by the Contractor but including any other matters or directions which under these Conditions are to be treated as, or as requiring, a Variation)'.
- Clause 4.20.2: 'Contractor's directions, including those which pass on the Architect/Contract Administrator's instructions:
  1 for the expenditure of Provisional Sums included in the Contractor's Requirements or any Bills of Quantities, excluding directions for expenditure of a Provisional Sum for defined work;
  2 for the opening up for inspection or testing of any work, materials or goods under clause 3.17 of the Main Contract Conditions (including making good), unless the inspection or test shows that the work, materials or goods were not in accordance with the Sub-contract;
  3 for the opening up for inspection or testing of any work, materials or goods under clause 3.10 (including making good), unless the inspection or test shows that the work, materials or goods were not in accordance with this Sub-contract;
  4 in relation to any discrepancy in or divergence between any of the Numbered Documents or any discrepancy in or divergence between any of those documents and the Contract Documents under the Main Contract;
  5 for the postponement of any work to be executed under this Sub-contract (whether in connection with a postponement under the Main Contract or otherwise)'.
- Clause 4.20.3: 'compliance with clause 3.22.1 of the Main Contract Conditions and related directions, including those passing on instructions under clause 3.22.2 of those Conditions'.
- Clause 4.20.4: 'Suspension by the Contractor under clause 4.14 of the Main Contract Conditions of the performance of his obligations under the Main Contract'.
- Clause 4.20.5: 'the execution of work for which an Approximate Quantity included in any Contract Bills is not a reasonably accurate forecast of the quantity of work required'.
- Clause 4.20.6: 'where there are Bills of Quantities, the execution of work for which an Approximate Quantity included in those bills is not a reasonably accurate forecast of the quantity of work required'.
- Clause 4.20.7: 'Any impediment, prevention or default, whether by act or by omission, by the Employer, the Architect/Contract Administrator, the Quantity Surveyor or any of the Employer's Persons, except to the extent caused or contributed to by any default, whether by act or omission, of the Sub-contractor or of any of the Sub-contractor's Persons'.

- Clause 4.20.8: 'Any impediment, prevention or default, whether by act or by omission, by the Contractor or any of the Contractor's Persons (including, where the Contractor is the Principal Contractor, any default, whether by act or omission, in that capacity) except to the extent caused or contributed to by any default, whether by act or omission, of the Sub-contractor or of any of the Sub-contractor's Persons'.

In respect of the words: 'except to the extent caused or contributed to by any default, whether by act or omission, of the Sub-contractor or of any of the Sub-contractor's Persons', within clauses 4.20.7 and 4.20.8 of the sub-contract, it should be particularly noted that clauses 2.7.3 and 2.7.4 of the sub-contract require that:

- such further drawings, details, information and direction referred to in clauses 2.7.1 and 2.7.2 of the sub-contract shall be provided or given at the time it is reasonably necessary for the recipient party to receive them, having regard to the progress of the sub-contract works and the main contract works;
- where the recipient party has reason to believe that the other party is not aware of the time by which the recipient needs to receive such further drawings, details, information or directions, he or she shall, so far as is reasonably practicable, advise the other party sufficiently in advance to enable him or her to comply with the requirements of clause 2.7 of the sub-contract.

9.6   Can the contractor recover direct loss and/or expense from the sub-contractor?

Clause 4.21 gives the contractor a right to recover from the sub-contractor the agreed amount of any direct loss and expense caused to the contractor where the 'regular progress of the main contract works' is materially affected by any act, omission or default of the sub-contractor, or any of the sub-contractor's persons.

Clearly the contractor has the burden of proof in demonstrating that the above qualifying criteria exists in order for this provision to apply. This may be problematic in practice, especially if there are other potentially competing causes of delay and/or disruption caused by the contractor and/or his or her other sub-contractors.

Clause 2.21 also deals with the recovery by the contractor of any direct loss and/or expense suffered or incurred by the contractor caused by the culpable failure of the sub-contractor to complete the sub-contract works on time.

In respect of clause 4.21, the contractor is obliged to notify the sub-contractor in writing with reasonable particulars of the effects or likely effects on the regular progress of the main contract works resulting from any act, omission or default of the sub-contractor, or any of the sub-contractor's persons, and shall also give details of the resultant loss and expense as the sub-contractor reasonably requests.

Any amount of loss and expense agreed between the contractor and the sub-contractor relating to clause 4.21.1 may be deducted from any monies due to or to become due to the sub-contractor or shall be recoverable by the contractor from the sub-contractor as a debt.

Whilst not normally specifically relevant to a contractor–sub-contractor relationship, and whilst not catered for within the JCT sub-contracts under consideration within this book, a brief note should be made of liquidated damages.

The parties to a contract often agree that a liquidated (i.e. fixed and agreed) sum shall be paid as damages for some breach of a contract.

A typical clause provides that if a contractor fails to complete by a date stipulated in the contract, or by any extended date, he or she shall pay or allow the employer to deduct liquidated damages at the rate of a certain amount per week for the period during which the works are uncompleted.

The basic rules in respect of liquidated damages are that they cannot be a 'penalty' and they must be a genuine pre-estimate of the level of damages that would be incurred.

It should be noted that a contractor may be able to recover from a sub-contractor the liquidated damages that he or she has incurred as part of a common law damages claim (or, indeed, as part of a loss and expense claim), provided that the sub-contractor was made aware of that level of liquidated damages at the time that the sub-contract was placed.

9.7   In pursuing a loss and expense claim, what does a sub-contractor need to prove?

Most claims, whether for time or money, involve establishing what is often called the nexus of cause and effect (i.e. the link between cause and effect). This means that what needs to be proved is that, because of the occurrence of a particular event, certain things happened, and as a direct result of those events happening, this in turn led to one of the parties incurring delays or costs which were not previously contemplated by the parties and which it would not have been reasonable for the parties to have contemplated.

In order to establish this nexus (or link), which is never an easy task in a construction situation, records need to be available which show that the circumstances that existed before an event occurred changed after that event occurred, and that that change in circumstances could only have been as a result of the event in question.

9.7.1   Burden of proof

Ordinarily, the burden of proof lies with the party who makes an assertion; hence, the legal maxim 'he who asserts must prove'.

Accordingly, it is not for a party defending a claim to disprove the claim; it is for the party pursuing a claim to prove the claim.

However, if the party pursuing a claim adduces sufficient evidence to raise a presumption that what is claimed is true, the burden of proof will pass to the other party. It is then for that party to adduce sufficient evidence to rebut the presumption.

Therefore, any claim document should be prepared with the above basic principle in mind.

9.7.2   Standard of proof – what needs to be provided?

The standard of proof required in civil proceedings is referred to as the 'balance of probabilities' principle, whereby the court/tribunal makes its decision on the basis that something is more likely to have occurred than not.

Therefore, if a set of scales is imagined as being the balance of probabilities, there only needs to be 51% of probability on one side of the scales to enable that side of the scales to succeed over the other side of the scales. Naturally, it is necessary for the party with the 'burden of proof' at any particular time to push the scales down to (at least) the 51% level.

This is to be contrasted with the 'standard of proof' in criminal proceedings, which is set at a much higher level of 'beyond reasonable doubt'.

When preparing a claim (in addition to the burden of proof principle outlined above), the required standard of proof needs to be recognized, particularly as, in certain situations, the party receiving the claim (i.e. before proceedings are commenced) may appear to expect a standard of proof more akin to 'beyond reasonable doubt' rather than on the balance of probabilities.

Part of this reason may be because of the apparent dichotomy between the standard of proof required for civil proceedings (i.e. the balance of probabilities) and the requirement under most of the JCT sub-contracts (e.g. clause 4.19.3) which assumes that the contractor will *ascertain* the loss and/or expense value.

The word 'ascertain' is usually understood to mean 'to find out for certain'. However, this is much closer to the standard of proof for criminal proceedings (i.e. beyond reasonable doubt) than it is to the standard of proof for civil proceedings (i.e. the balance of probabilities).

There have been conflicting court cases regarding this particular matter. For example, in the case of *Alfred McAlpine* v. *Property and Land*,[2] it was found that when an architect (in that case) was required to ascertain, he or she is obliged to find out for certain and not merely to make a general assessment. However, in the *How* v. *Lindnor*[3] case, it was held that assessment of loss and expense was akin to assessment of damages, requiring no special standard of proof, and that the exercise of judgment was not only permissible but required where loss had not been proved with absolute certainty.

Further, in the case of *Norwest Holst* v. *Co-op*,[4] Judge Thornton found that under a sub-contract, the parties were normally required to agree loss and

---

[2]*Alfred McAlpine Homes North Ltd* v. *Property and Land Contractors Ltd* (1995) 76 BLR 59.
[3]*How Engineering Services Ltd* v. *Lindnor Ceilings Ltd[1999] CILL 1521.*
[4]*Norwest Holst Construction* v. *Co-op Wholesale Society* (1997/1998) (unreported).

expense, and in the event that they could not agree, it was in order for an arbitrator to determine such reasonable loss and expense (on a balance of probabilities basis).

Another point to note is that when assessing probabilities, the court or tribunal will have in mind, to whatever extent is appropriate in the particular case, that the more serious the allegation the less likely it is that the event occurred and, hence, the stronger should be the evidence before the court concludes that the allegation is established on the balance of probabilities.

This does not mean that different standards of proof are required where different assertions are made, but it does mean that the inherent probability or improbability of an event is itself a matter to be taken into account when weighing the probabilities and deciding whether, on balance, the event occurred.

The more improbable the event, the stronger must be the evidence that it did occur before, on the balance of probability, its occurrence will be established.[5]

It should be noted that 'more likely than not' is not necessarily a hard test to pass and facts can often be proved to this standard merely by circumstantial evidence, or by one person's word being preferred to another's, although contemporaneous written records are always more influential.

In all circumstances, some admissible evidence must be produced to support everything claimed.

9.8   What needs to be proved in a loss and expense claim document?

Following on from the above, the initial points that need to be proved in a loss and expense claim document are:

■   that an event actually occurred;
■   that the event was one expressly catered for within the sub-contract;
■   that the notices required under the sub-contract had been given;
■   what the effect was, in financial terms, of the specified event.

In respect of the above list, what then needs to be shown is the connection between the *cause* and the *effect*, that is, showing how the latter item on the above list (the financial effect) resulted directly from the first item on the above list (the event that occurred).

This is the part of the process that normally causes most difficulty.

The above connection is a matter of evidence and it is the part of a claim for which records are absolutely crucial.

Indeed, this is when the quality and extent of the records and documentation come into their own.

With some of the heads of claim, such a connection is relatively easy to identify; whilst with others, it is rather more problematic.

---

[5]*Ref Lord Nicholls in Re H* (Minors) [1996] AC 563.

It is because of the difficulty of establishing the required nexus of cause and effect that it is not uncommon for global claims to be produced.

Before considering the individual heads of claim in detail, it is worth looking at the background of global claims generally.

9.9   What is a global claim?

A global claim is a claim for financial loss which arises from various different events; however, the requirement to link cause and effect is absent (i.e. individual sums of money are not claimed for each individual event). Instead a single global sum is claimed in respect of the alleged cumulative effect of all of the events.

Quite often claimants will simply base their claims in terms of quantum on the difference between estimated cost and actual cost, normally with an adjustment for any recovery made within the variation account.

A simple example is as follows:

Total costs incurred by the sub-contractor	£1,000,000
Less: sub-contract order value	£ 600,000
Less: recovery from variations	£ 200,000
Claim for loss and/or expense (i.e. the shortfall)	£ 200,000

The basis of the global claim being advanced is that the total extra cost incurred, as calculated above, has resulted from numerous events whose consequences had such a complex interaction that it is impossible or impractical to disentangle them to show cause and effect.

Unsurprisingly, global claims have been the subject of much controversy and have been considered in many court cases, remaining a difficult and developing area of law.

The 2012 case of *Walter Lilly* v. *DMW Developments*[6] provided useful guidance in respect of global claims. In that case, Mr Justice Akenhead concluded that in respect of global claims:

(a)   claims for delay- or disruption-related loss and expense must be proved as a matter of fact. Thus, it must be demonstrated on a balance of probabilities basis that, first, events occurred, which entitle loss and expense to be paid; second, that those events caused delay and/or disruption; and third, that such delay or disruption caused loss and/or expense (or loss and damage as the case may be) to be incurred.

(b)   it is open to contractors and sub-contractors to prove these above three elements with whatever evidence that will satisfy the tribunal and the requisite standard of proof. There is no set way for contractors or sub-contractors to prove these three elements. For instance, such a claim

---

[6] *Walter Lilly & Company Ltd* v. *DMW Developments Ltd[2012] EWHC 1773 (TCC).*

may be supported or even established by admission evidence or by detailed factual evidence which precisely links reimbursable events with individual days or weeks of delay or with individual instances of disruption and which then demonstrate with precision to the nearest penny what that delay or disruption actually cost.

(c)   There is nothing, in principle, 'wrong' with a 'total' or 'global' cost claim. However, there are added evidential difficulties (in many but not necessarily all cases) which a claimant has to overcome. A claimant will generally have to establish (on a balance of probabilities) that the loss which he or she has incurred (namely the difference between what he or she has cost and what he or she has been paid) would not have been incurred in any event. Thus, he or she will need to demonstrate that his or her accepted tender was sufficiently well priced that he or she would have made some net return. He or she will need to demonstrate in effect that there are no other matters which actually occurred (other than those relied upon in his or her pleaded case *and* which he or she has proved are likely to have caused the loss).

(d)   The fact that one or a series of events or factors which are the risk or fault of the claimant caused or contributed (or cannot be proved not to have caused or contributed) to the total or global loss does not necessarily mean that the claimant can recover nothing. It depends on what the impact of those events or factors is.

(e)   Obviously, there is no need for the court to go down the global or total cost route if the actual cost attributable to individual loss-causing events can be readily or practicably determined, but a claimant is not debarred from pursuing a global claim if he or she could otherwise seek to prove his or her loss in another way. It may be that the Court or the tribunal will be more sceptical about the global cost claim if the direct linkage approach is readily available but is not deployed, but that does not mean that the global cost claim approach should be rejected out of hand.

Generally, the best practice to follow in preparing a claim involving multiple variations, delay, disruption and extra cost is to follow the guidelines laid down by the courts strictly, so far as it is possible, and to present a cause and effect claim where possible, and a global claim only in respect of any element of the claim where it is simply not possible or practical to do otherwise.

One way of presenting this type of claim is to have a points of claim document which deals with the legal issues, setting out the terms of the contract which are applicable, in particular the variations provisions, any terms which give rise to a claim for breach of contract, and setting out in summary form in what respects the breaches occurred. This document would be relatively short and the detailed factual matters supporting the claim would be set out in a separate document.

The separate document could be in the form of a 'cause and effect' schedule which, for a claim based mainly on variations, might be set out as follows:

- Column 1 – Event
- Column 2 – Clause No.

- Column 3 – Other Term
- Column 4 – Effect (Disruption)
- Column 5 – Delay (From and To)
- Column 6 – Claim/Loss (£)

where:

- Column 1 would set out the event relied on, for example, a variation order or an event which is said to be a breach of the sub-contract.
- Column 2 would simply refer to an appropriate and relevant sub-contract clause number or, if the clause has sub-clauses, the appropriate and relevant sub-contract sub-clause number.
- Column 3 would be used if some other express or implied terms were relied upon for events other than variations.
- Column 4 would be used to demonstrate the effect. This would most likely be a narrative if the claim related to disruption, as opposed to a prolongation claim, explaining what activity was affected by the event complained of and in what way.
- Column 5 would deal with prolongation, giving the precise start date of delay and the precise finish date of delay. It would also be appropriate to indicate in that column whether delay was continuous or intermittent, since it is a common failing in this type of claim to adopt a notional rather than a historical analysis in which it is assumed that where something goes wrong, a continuous delay to the contract as a whole occurs until that matter is put right, which is often not the case. It is also necessary in assessing prolongation to include an analysis of concurrent delay, so that critical delay is identified (another common failing in this type of claim being to assume that all delay is cumulative or failing to identify what is concurrent and what is cumulative).
- Column 6 would include an amount which is claimable under the terms of the contract, for each individual item noted in column 1.

In attempting to produce this type of schedule, it will be found in some cases that it is not possible with multiple variations, and possibly other claims, to give a precise delay period or claim figure for each one, because of the complex interaction of events. In that event, all one can do is to break it down *as far as possible* and to present the remainder as a global claim.

Approaching the matter in this way is not only a good discipline but also more likely to result in success, as it closely replicates the latest requirements of the courts.

If faced with a global claim, it is fairly standard to request a 'cause and effect' analysis as this will highlight the weaknesses of the other party's case.

9.10 Common heads of a loss and expense claim

The common heads of a loss and expense claim are described as follows:

- Prolongation costs
- Disruption claims

- Winter working
- Head office overheads and profit
- Loss of profit
- Increased costs
- Cost of claim preparation
- Interest and finance charges
- Acceleration
- Common law damages

9.11   Prolongation costs

These are also commonly referred to as 'preliminary items'. Additional expense under this heading is usually easily ascertainable from the sub-contractor's cost records.

Such costs may include the following examples:

- Project manager (site based)
- Site foreman
- Site quantity surveyor
- Site engineer and assistants
- Site offices
- Site administration costs
- Site stores
- Welfare facilities (if provided by the sub-contractor)
- Cranage/Forklifts
- General and small plant
- Scaffolding/Towers
- Telephone
- Electricity
- Insurance

What the sub-contractor is seeking to establish is that, because of events that have occurred and for which the sub-contract provides an entitlement, actual resources of the type listed above (amongst others), additional to those which otherwise would have been necessary, have had quite reasonably to be engaged on the project.

9.11.1   What is the basis of the claim under this heading?

The sub-contractor is essentially seeking to recover site overhead costs, that is, those costs which are not related to one specific item of work but which are necessary for the site establishment generally.

The site overheads part of the claim is commonly broken down into delay costs (due to prolongation) and 'thickening' (or extra resource) costs (due to disruption).

The claim is usually for time-related costs, that is, the additional cost associated with the prolonged use of a particular resource; for example, a 5-week delay will clearly affect the sub-contractor's time-related costs, such as site supervision, cabins, hired equipment, etc.

However, this may not always be the case. It may be that the sub-contractor incurs *additional* supervision costs without the overall completion date for the sub-contract works being affected. For example, *additional* supervision costs may be incurred by the sub-contractor during the original sub- contract period, or a sub-contractor may be delayed or disrupted by an instruction ordering additional work to non-critical items (e.g. scaffold to an area of work which was not on the critical path). In such circumstances, it may be entitled to loss and expense incurred (i.e. thickening as noted above) regardless of whether the additional work delayed the sub-contractor in the completion of the works.

9.11.2   Are the delay costs (prolongations) taken as those in the over-run period?

The prolongation part of the claim must relate to those periods when delay occurred; this will not necessarily be the overrun period.

For example, if a sub-contractor had a sub-contract period of 6 weeks, but was delayed in week 2 for 1 week, the site overhead costs would need to be claimed for week 2 (i.e. the week when the delay occurred) and not for week 7 (i.e. the week when the delay effect became apparent).

9.11.3   What costs are claimed?

The actual costs incurred by the sub-contractor are to be claimed. A pro-rata of bill/tendered allowances does not represent actual cost and is therefore likely not to be successful.

9.11.4   Is a sub-contractor entitled to additional site overhead costs where he or she programmes to complete a project before the contract completion date but is prevented from doing so?

There is much general debate surrounding early finish programmes.

The sub-contract on-site period may be, for example, eight weeks, but the sub-contractor may decide that he or she wishes to complete the sub-contract works in 6 weeks. However, the contractor issues instructions that prevent the sub-contractor from completing the sub-contract works until the end of week 8. The sub-contractor then wishes to recover the additional site overhead costs for weeks 7 and 8.

In the *Glenlion Construction* v. *The Guinness Trust*[7] case, the position in respect of JCT contracts was made reasonably clear. In that case (which related to a contractor–employer relationship), it was found that the contractor was not entitled to an extension of time in such a situation. Under JCT contracts, extensions of time are only due if completion of the works is delayed beyond the completion date.

---

[7]*Glenlion Construction Ltd* v. *The Guinness Trust(1987) 39 BLR 89.*

Despite the above, if a sub-contractor can show that he or she has been caused loss and expense by reason of the delay, he or she may be entitled to claim for loss and expense.

9.11.5 What points need to be considered in assessing such a claim?

In assessing such a claim several points should be kept in mind:

- What has to be established in the first instance is that additional resources were required.
- The sub-contractor needs to establish with sufficient particularity:
  — what was included in his or her tender;
  — that the tender allowance was sufficient.
- Just because resources are engaged during a period of overrun does not mean they are additional. For example, if a sub-contractor originally expected to have two foremen on site for 10 weeks but actually only had one foreman on site for 20 weeks, there would not (on the face of it) be any additional costs.
- The sub-contractor needs to establish that the additional resources were required due to the reasons that he or she is relying on.
  This may be done in several ways:
  o If the claim is for prolongation, then it is necessary to show that resources were retained on site during the periods of delay for reasons outside of the subcontractor's control. This is usually one of the easier points of the claim.
       However, the sub-contractor does not need to have received an extension of time before he or she can pursue such prolongation costs during a period in excess of the original period.
       Conversely, not all extensions of time circumstances give rise to financial entitlement, and even if they do, not all the additional resources may be recoverable.
  o A claim for site overhead costs may be pursued in the absence of prolongation.
       If an appropriate resource is required for longer during the original sub-contract period, or more of them are required for the same period, then they should be recovered if they can be associated with a specific event.
       This may apply, for example, in the situation where a variation is issued which results in certain work activities which were programmed to be executed consecutively now needing to be carried out concurrently, thus requiring two or more work faces to be operating and needing two (or more) sets of supervisors, etc., where previously one would have been adequate.
- The sub-contractor then has to establish the cost of the resources.
  With supervisory staff, this would be ascertained using salary levels, pensions, national insurance, car costs, petrol, bonuses, etc. If agency staff have been used, then their actual costs will be recoverable providing that it was reasonable to use them.

9.12   Disruption claims

Delay and disruption can lead to increased expenditure on plant and labour in two ways. One is that it may be necessary to employ extra resources, and the other is that existing resources may be used inefficiently.

This head of claim in particular may be difficult to ascertain due to lack of records and the difficulty of establishing the nexus of cause and effect.

However, just because it is difficult to establish the above, it does not mean that the sub-contractor should not submit a claim. Equally, it does not mean that the contractor should not make a reasonable assessment.

The 'global' claim approach should be used very much as the exception rather than the rule. It is applicable and permitted only where the events are so numerous/complex/interrelated that no amount of analysis or record keeping would establish or evidence the connection of cause and effect on individual matters. However, it is not good enough that such analysis cannot be done simply because the sub-contractor has not kept reasonable records.

There are a number of methods of assessing this head of claim, and indeed few people agree on the approach to use. However, in all cases, good records are essential.

Ideally, what should be established is that, *before* a particular event occurred, productive resources were achieving a certain level of productivity/financial income for the sub-contractor. However, *after* that particular event occurred and indeed, because of it, the same level of productivity/financial income could not be achieved.

Take for example the case of a bricklaying sub-contractor.

In the sub-contract, the sub-contractor is told that he or she will be required to build wall A, followed by wall B, and followed by wall C. In his or her tender, the sub-contractor allowed for an output of 60 bricks per hour per bricklayer:

- Week 1: Things progress well. The bricklayers worked on wall A only, as planned; the output achieved was 65 bricks per hour. The sub-contractor was exceeding his or her tender expectations and this illustrated the sufficiency of his or her tender.
- Week 2: Things start to go wrong. The sequence of work changed, and the sub-contractor was instructed by the contractor to switch his or her bricklayers from wall A to wall B to wall C, often all in the same day. As a consequence, the output of the bricklayers at the end of the second week reduced to 50 bricks per hour.

In such a case, the disruption claim would be based on the loss of production caused by the disruption event. In other words, the sub-contractor expected to achieve 60 bricks per hour (or the sub-contractor might argue that he or she actually achieved 65 bricks per hour) before the disruption event occurred, but because of the disruption event, the sub-contractor's bricklayers only achieved 50 bricks per hour.

Although the above is a very simple example, the principle remains the same for much more complicated claims. Clearly, for this head of claim, more than any other, good, complete and accurate records are absolutely crucial.

There are several types of approach in respect of a disruption claim which may be applicable, depending on the nature of the work, the circumstances and the records available. The main approaches used being:

- To show that resources were standing idle when they should otherwise have been working – thus taking longer to do the same work
- To illustrate how the work of a particular trade has been prolonged and to apply a disruption factor in the same proportions to the cost of the trade
- To estimate a disruption factor
- To do an overall global claim based on costs of all trades assessed in the tender and to claim the difference between that and actual costs
- To do a mini global claim using the approach as last but targeting specific trades
- To illustrate the resources employed in producing specific and evidenced amount of work prior to an event and to claim the extra resources for doing equivalent work after the event
- To show that more expensive resources were required to do the same value of work because of an event, even though the level of resources was the same, that is, agency resources, emergency hire of plant, expensive gangs, etc.

9.13   Winter working

This heading amounts to a different aspect of loss of productivity resulting from the need to work in less favourable climatic conditions. This is a recoverable head of claim.[8]

What the sub-contractor has to establish in this case is that for reasons providing an express contractual entitlement, he or she has had to undertake productive work in climatic conditions less favourable than would otherwise have been the case.

It is not enough to show only that work was undertaken during the winter (for example); it must also be shown that this fact did actually affect the work.

Of course, this type of claim could be problematic for sub-contractors, particularly if their actual period on site could have commenced at any time, within wide parameters.

9.14   Head office overheads and profit

Head office overheads are sometimes called 'off-site overheads' or 'establishment charges'.

Head office overheads are those administrative and management costs of running the head office of a sub-contractor's business over and above the site costs.

[8]*Bush* v. *Whitehaven Port & Town* (1888) 52 JP 392.

Generally, head office overheads will include for:

- purchase and/or rent of office and yards
- maintenance and running costs
- transport and mobilisation costs
- company cars
- depreciation
- Director's emoluments and expenses
- salary and other cost of staff
- administration costs
- legal and audit fees, etc.

9.14.1 What is the basis of the claim of additional head office overheads costs?

Most sub-contractor's head office resources exist to support operations undertaken on site, and such head office overheads are normally taken as being recovered out of the income from his or her business as a whole.

Where completion of one project has been delayed, a sub-contractor may claim to have suffered a loss arising from the diminution of his or her income from the project reducing the turnover of his or her business. Despite this, the sub-contractor continues to incur expenditure on head office overheads which he or she cannot materially reduce or, in respect of the project, can only reduce, if at all, to a limited extent.

Therefore, but for the delay and disruption, the sub-contractor's workforce would have had the opportunity of being employed on another project, which would have had the effect of contributing to the head office overheads and profit during the overrun period.

There is some authority that a claim on this basis is sustainable.

In *JF Finnegan* v. *Sheffield City Council*,[9] Judge Sir William Stabb (talking about a contractor in that case) said:

> 'It is generally accepted that, on principle, a contractor who is delayed in completing a contract due to the default of his employer, may properly have a claim for head office or off-site overheads during the period of delay, on the basis that the workforce, but for the delay, might have had the opportunity of being employed on another contract which would have had the effect of funding the overheads during the overrun period'.

It should be noted that the entitlement to such general overheads is in itself an arguable point. However, the consensus of opinion is in favour, provided, of course, it can be established that such overheads would have been recovered were it not for the delay.

However, substantial claims of this kind are rarely made because most sub-contractors are able to cope with delay on a particular contract using their existing resources whose cost is reasonably constant.

---

[9] *JF Finnegan Ltd* v. *Sheffield City Council* (1988) 43 BLR 124.

9.14.2   What must the sub-contractor seek to demonstrate?

In terms of first establishing liability, the sub-contractor must seek to prove that head office overhead costs have been increased as a result of a delay and disruption suffered by the sub-contractor, for which he or she is entitled to recompense under the sub-contract. Examples may range from the cost of extra staff recruited because the particular project was in difficulties to the cost of extra telephone calls and postage in the period of delay.

It is suggested that, in order to succeed, a sub-contractor must provide evidence to show:

(1)   that the profit or overhead contribution was capable of being earned elsewhere at the time of delay;
(2)   that the profit and overhead percentage is a reasonable one;
(3)   that work of the same level of profitability and/or overhead recovery was available during the period of delay.

In terms of evidencing that there was other work available which, but for the delay, he or she would have secured, the sub-contractor might seek to evidence this by producing declined invitations to tender, with evidence that the reason for declining was that the delay in question left him or her insufficient capacity to undertake other work.

He or she might alternatively show from his or her accounts a drop in turnover and establish that this resulted from the particular delay rather than from extraneous causes. If loss of turnover resulting from delay is not established, the effect of the delay is only that receipt of the money is delayed. It is not lost.

If liability is established, then the sub-contractor must establish quantum. When pursuing a claim for head office overheads and profit, a sub-contractor will frequently rely on the use of a formula to evidence such quantum.

9.14.3   What formulae are used?

There are three formulae that are in common usage (and, in addition, there are many other derivatives from these formulae).

These formulae are:

■   the Hudson formula;
■   the Emden's formula;
■   the Eichleay formula.

Whilst the use of formulae has been heavily criticised by many commentators, there is judicial authority for their use in certain instances. However, whatever formula may be used, it must be noted that the use of a formula merely provides the means for assessing quantum. The fact that there was an actual loss, and that this loss flows directly from the relevant matter relied upon must first be established.

In addition, when submitting claims for head office overheads and profit, care needs to be taken to avoid any double recovery or overlap with other claims or payments obtained by the sub-contractor, such as variations which have been computed by using the contract prices as a basis. Normally, in such a situation, the valuation of the variation will include for an element of additional head office overheads and profit recovery.

9.15   Loss of profit

Although there has been some debate about whether loss of profit is recoverable as part of a loss and expense claim, if the words 'direct loss and expense' are to be interpreted as equivalent to the measure of damages at common law for breach of contract (which is generally accepted to be the case), then there can really be no doubt that a sub-contractor can claim for the loss of profit that he or she would have earned on other sub-contracts had there been no delay and disruption to the project in question, provided that such loss of profit was foreseeable.

Clearly, for a claim of loss of profit to be made, a sub-contractor would need to show, as with head office overheads, that at the time of the delay, he or she could have used the lost turnover profitably. A claim for loss of profit does not, it is submitted, fail merely because the contract in question was unprofitable. The question is what the contractor would have done with the money if he or she had received it at the proper time. Even if, at that time, the contractor's business was making a loss, a sum analogous to loss of profit is, it is submitted, recoverable if the loss of turnover increased the loss of business.

9.16   Increased costs

This head of claim usually arises on a fixed-price sub-contract (although it can equally apply where a fluctuation recovery has been agreed) where expenditure on resources due to increased costs during an extended period is sought.

It is submitted that the correct measure is the difference between what the sub-contractor would have spent on resources and what he or she has actually spent due to delays suffered. There are, however, a number of different methods for calculating this head of claim.

The best way is for the claim to be based on substantiated details of the level of costs that would have been expended on resources and the actual costs paid for those same resources backed up by invoices, etc. However, it is rare in practical terms that such an analysis could be economically undertaken.

Therefore, an alternative method would be to use some type of formula approach.

Using a formula approach, the calculation would broadly be as follows:

- Ascertain the value of resources in the original tender but using actual work values.
- Find out the level of inflation for such resources over the original sub-contract period, normally by using some standard published indices.

- Calculate how much the cost of resources would have increased during the original sub-contract period based on the indices obtained. This will then be taken as the amount deemed to be allowed in the rates.
- Undertake the same calculation but using the indexed inflation figure over the extended sub-contract period.

The extra over cost between the above two calculations will constitute the increased cost claim.

9.17 Cost of claim preparation

Where the preparation of a claim is undertaken by a sub-contractor's in-house staff, this cost will often be partly (or completely) reflected in the sub-contractor's loss and expense claim for site overheads or head office overheads.

Often, a sub-contractor may employ a consultant to assist in the preparation of a loss and expense claim, and the general rule is that the costs incurred in taking this action are generally not recoverable, except and until the matter proceeds to arbitration[10] or litigation.

The reasoning behind disallowing this head of claim is that the sub-contractor is obliged, under the sub-contract, to provide information, etc. to resolve financial issues. Therefore, the submission of a claim and the work involved must have been included in his or her price.

It may be the case, however, that such monies might be recovered if the contractor requested further information from the sub-contractor which is judged to be beyond the sub-contract requirements.

9.18 Interest and finance charges

When a sub-contractor incurs loss and expense, this has to be financed by him or her either from his or her own capital resources or, alternatively, by increased borrowing. In either case, it is clear that the use of money costs money.

Depending on the level of inflation (and the interest rates applicable at any point in time), and the amount outstanding in respect of the loss and expense claim, this matter will either be a relatively small matter or will have major implications. In either event, it is a matter of great importance to the sub-contractor how such interest or finance charges should be recovered.

Interest in respect of monies owed is dealt with as an express term of the sub-contracts. Therefore, if it is proven that monies were outstanding in respect of the non-payment or under-payment of a loss and expense claim, simple interest would be applied in line with the sub-contract in question.

In the alternative case, finance charges, or the cost of being stood out of one's money, are a recoverable head of a loss and expense claim in any event. This

---

[10]*James Longley & Co Ltd v. South West Thames Regional Health Authority* (1984) 25 BLR 56.

was established beyond doubt in the Court of Appeal case of *FG Minter Ltd* v. *WHTSO*.[11]

In respect of common law damages, to recover finance charges, it is necessary for the claimant to show that the finance charges fell within the second limb of *Hadley* v. *Baxendale*, that is, that the finance charges are special damages that were in the contemplation of the parties at the time that the contract was formed.[12]

9.19 Acceleration

Acceleration may best be explained by considering a typical example. A delay is caused by a failing of the contractor. Mindful of the financial consequences on the contractor of finishing late (e.g. liquidated damages and prolongation costs), the contractor may be more interested in the sub-contractor completing his or herworks in line with the original dates rather than in extending the sub-contractor's period of time on site.

In such a situation, it is not uncommon for a contractor to attempt to reach some form of acceleration agreement with the sub-contractor, whereby the sub-contractor will take measures (e.g. weekend working, double shifts, increased resources) to complete 'on time' (or, at least, earlier than would have been achieved under the normal course of events), and if the earlier required date is achieved, the contractor will make an agreed acceleration payment to the sub-contractor.

In the case of delay, the only express action available to the contractor under the sub-contract is to extend the sub-contractor's period of time on site.

It is debatable whether a schedule 2 quotation provided by a sub-contractor incorporates the possibility of agreeing acceleration to the sub-contract.

On the face of it, this would be possible because clause 2.2 of the said schedule 2 quotation states that the quotation shall comprise 'any adjustment required to the time period'. This 'adjustment' could naturally include for both an increase and a decrease to the previously agreed time period. A decrease to the previously agreed time period could be as a result of acceleration measures, and those acceleration measures could be priced within a schedule 2 quotation.

The reason why the word 'debatable' is used above is that clause 2.2 of the schedule 2 quotation for the main contract specifically states 'including, where relevant, stating an earlier Completion Date than the Date for Completion given in the Contract Particulars', but no such similar wording is included in clause 2.2 of the schedule 2 quotation document within the sub-contract.

Nevertheless, this is a matter for the parties to agree, and it is therefore questionable whether the omission of the words in question has any real practical effect.

Apart from the above possibility, the sub-contract does not specifically allow for acceleration agreements, and great care needs to be taken by both the contractor and the sub-contractor when entering into any such agreement.

---

[11]*FG Minter Ltd* v. *WHTSO* (1980) 13 BLR 1, CA.

[12]*Robinson* v. *Harman* (1848) 1 Ex 850.

A sub-contractor should not undertake acceleration measures (unless those measures are in mitigation of delays that the sub-contractor has caused) without reaching some form of agreement with the contractor, otherwise the probability is that he or she will not receive any payment for the acceleration measures undertaken.

9.20   Common law damages

Clause 4.22 preserves the common law rights of the contractor and the sub-contractor.

In respect of these common law rights, a breach of contract which has not been excused gives the injured party the right to bring an action for damages.

9.20.1   What are damages?

There are many classic definitions but, essentially, damages are awarded 'so far as money can do it'[12] to put the claimant as nearly as possible 'in the same position as he would have been in if he had not sustained the wrong for which he is now getting compensation or reparation'.[13]

However, the courts set a limit to the loss for which damages are recoverable, and loss beyond such limit is said to be too remote.

In this regard, the rule of remoteness is set out in the case of *Hadley* v. *Baxendale*,[14] and this is:

'Where two parties have made a contract which one of them has broken, the damages which the other party ought to receive in respect of such breach of contract should be such as may fairly and reasonably be considered either:

(1)   arising naturally, i.e. according to the usual course of things from such breach of contract itself; or
(2)   such as may reasonably be supposed to have been in the contemplation of both parties at the time they made the contract, as the probable result of the breach of it'.

The rule as stated above is recognised as having two limbs, and these limbs have been indicated in the above quoted text by the insertion of the numbers (1) and (2).

9.20.2   The first limb of *Hadley v. Baxendale*

There have been two leading cases where the *Hadley* v. *Baxendale* rule has been considered. These are the *Victoria Laundry* v. *Newman Industries*[15] case and the *Koufos* v. *Czarnikow*[16] case.

---

[13] *Lord Blackburn in Livingstone v. Rawyards Coal Company* (1880) 5 App Cas 25, HL.
[14] *Hadley* v. *Baxendale* (1854) 9 Ex 341.
[15] *Victoria Laundry (Windsor) Ltd* v. *Newman Industries Ltd[1949] 2 KB 528* .
[16] *Koufos* v. *Czarnikow[1969] 1 AC 350.*

In the *Victoria Laundry* case, Lord Justice Asquith stated several propositions that emerged, he said, from the authorities as a whole. Taking these into account together with the opinions of the House of Lords in the *Koufos* v. *Czarnikow* case, and other following cases, the first limb of the *Hadley* v. *Baxendale* rule may be elaborated into the following propositions:

(1) 'The aggrieved party is only entitled to recover such part of the loss actually resulting as may fairly and reasonably be considered as arising naturally, that is, according to the usual course of things, from the breach of contract.

(2) The question is to be judged as at the time of the contract.

(3) In order to make the contract breaker liable, it is not necessary that he should actually have asked himself what loss was liable to result from a breach of the kind which subsequently occurred. It suffices that, if he had considered the question, he would as a reasonable man have concluded that the loss of the type in question, not necessarily the specific loss, was "liable to result".

(4) The words "liable to result" should be read in the sense conveyed by the expressions "a serious possibility" and "a real danger" and "not unlikely to occur"'.

For the first limb, therefore, knowledge of certain basic facts according to the usual course of things is imputed, but not special knowledge.

9.20.3 The second limb of *Hadley v. Baxendale*

The second limb of *Hadley* v. **Baxendale** depends on additional special knowledge by the defendant.

The passage from *Hadley* v. *Baxendale* quoted above is followed by:

'If the special circumstances were communicated by the plaintiffs to the defendants, and thus known to both parties, the damages resulting from the breach of such a contract, which they would reasonably contemplate, would be the amount of injury which would ordinarily follow from a breach of contract under these special circumstances so known and communicated'.

As with the first limb, the question is to be judged at the time of the contract so that damages claimed under the second limb will not be awarded unless the claimant has particular evidence to show that the defendant then knew the special circumstances relied on.

9.20.4 Date of assessment

The general rule, both in contract and in tort, is that damages should be assessed as at the date when the cause of action arises.[17] But there are many exceptions,

---

[17] *Miliangos* v. *George Frank (Textiles) Ltd* [1976] AC 443, HL; *Dodd Properties* v. *Canterbury City Council* [1980] 1 WLR 433, CA.

and it has been said that 'this so-called general rule . . . has been so far eroded in recent times . . . that little of practical reality remains of it'.[18]

Thus, 'where it is necessary in order adequately to compensate the plaintiff for the damage suffered by reason of the defendant's wrong a different date of assessment can be selected'.[19]

9.20.5 Measure of damages

The purpose of damages is to put the innocent party (i.e. the claimant), 'so far as money can do it',[20] back to the same position as if the contract had been properly performed.

However, sometimes the proper measure of damages is not the cost of reinstatement but the difference in value between the work actually produced and the work that should have been produced.[21] This will particularly be so where the claimant has no prospect or intention of rebuilding, or where it would be unreasonable to award the cost of reinstatement.

The frequently quoted *Ruxley* v. *Forsyth*[22] House of Lords case was regarding a swimming pool that was not constructed to the required depth.

The swimming pool should have been constructed to a depth of 7 feet 6 inches; however, when the pool was completed, its depth was only 6 feet 9 inches.

Forsyth sought damages for the reinstatement costs (i.e. to re-build the swimming pool so that it was 7 feet 6 inches deep). However, the House of Lords found that where it would be unreasonable for the claimant to insist on reinstatement because the cost of the work involved would be out of all proportion to the benefit obtained, the claimant's measure of damages would simply be the difference in value (sometimes referred to as the loss of amenity value).

9.20.6 Mitigation of loss

The award of damages as compensation is qualified by a principle, 'which imposes on a plaintiff the duty of taking all reasonable steps to mitigate the loss consequent on the breach, and debars him from claiming any part of the damage which is due to his neglect to take such steps'.[23] But this 'does not impose on the plaintiff an obligation to take any step which a reasonable and prudent man would not ordinarily take in the course of his business'.[24]

---

[18]*Lord Justice Ormrod in Cory & Son v. Wingate Investments* (1980) 17 BLR 104, CA.

[19]*Lord Browne-Wilkinson in Smith New Court Ltd v. Scrimgeour Vickers* [1997] AC 254, HL.

[20]*Robinson v. Harman Clause 2.7.4* (1848) 1 Ex 850.

[21]*Dodd Properties v. Canterbury City Council* [1980] 1 WLR 433, CA.

[22]*Ruxley Electronics and Construction Ltd v. Forsyth[1995] 3 WLR 118.*

[23]*Lord Haldane in British Westinghouse v. Underground Railways Co* [1912] AC 673; *Andros Springs (Owners) v. World Beauty (Owners)* [1969] 3 All ER 158, CA; *Sotiros Shipping v. Sameiet Solholt* [1983] 1 Lloyds Rep 605, CA; *Kaines v. Osterreichische* [1993] 2 Lloyds Rep 1, CA.

[24]*British Westinghouse v. Underground Railways Co* [1912] AC 673 referring to Lord Justice James in *Dunkirk Colliery Co v. Lever* (1878) 9 Ch D 20, CA.

9.21   Equivalent sub-contract provisions

In the table below, the equivalent sub-contract provisions to that within the SBCSub/D/A or SBCSub/D/C in the text above are listed for the SBCSub/A or SBCSub/C, DBSub/A or DBSub/C, ICSub/A or ICSub/C and ICSub/D/A or ICSub/D/C.

SBCSub/D/A or SBCSub/D/C	SBCSub/A or SBCSub/C	DBSub/A or DBSub/C	ICSub/A or ICSub/C	ICSub/D/A or ICSub/D/C
Schedule 2 Quotation	Schedule 2 Quotation	Schedule 2 Quotation	Not applicable to this Sub-contract as Schedule 2 Quotations are not provided for under this Sub-contract	Not applicable to this Sub-contract as Schedule 2 Quotations are not provided for under this Sub-contract
Clause 2.7	Clause 2.7	Clause 2.7	Clauses 2.5 and 2.6	Clauses 2.5 and 2.6
Clause 2.7.1	Clause 2.7.1	Clause 2.7.1	Clause 2.5.1	Clause 2.5.1
Clause 2.7.2	Clause 2.7.2	Clause 2.7.2	Clause 2.5.2	Clause 2.5.2
Clause 2.7.3	Clause 2.7.3	Clause 2.7.3	Clause 2.6.1	Clause 2.6.1
Clause 2.19	Clause 2.19	Clause 2.19	Clause 2.13	Clause 2.13
Clause 2.21	Clause 2.21	Clause 2.21	Clause 2.15	Clause 2.15
Clause 3.10	Clause 3.10	Clause 3.10	Clause 3.9	Clause 3.9
Clause 4.19	Clause 4.19	Clause 4.19	Clause 4.16	Clause 4.16
Clause 4.19.3	Clause 4.19.3	Clause 4.19.3	Clause 4.16.3	Clause 4.16.3
Clause 4.20	Clause 4.20	Clause 4.20	Clause 4.17	Clause 4.17
Clause 4.20.1	Clause 4.20.1	Clause 4.20.1	Clause 4.17.1	Clause 4.17.1
Clause 4.20.2	Clause 4.20.2	Clause 4.20.2	Clause 4.17.2	Clause 4.17.2
Clause 4.20.3	Clause 4.20.3	Clause 4.20.3	Not applicable to this Sub-contract as this clause relates to a Relevant Sub-contract Matter (which relates to the discovery of antiquities, etc. under the Main Contract) that is not provided for under this Sub-contract	Not applicable to this Sub-contract as this clause relates to a Relevant Sub-contract Matter (which relates to the discovery of antiquities, etc. under the Main Contract) that is not provided for under this Sub-contract
Clause 4.20.4	Clause 4.20.4	Clause 4.20.4	Clause 4.17.3	Clause 4.17.3
Clause 4.20.5	Clause 4.20.5	Clause 4.20.5	Clause 4.17.4	Clause 4.17.4
Clause 4.20.6	Clause 4.20.6	Clause 4.20.6	Clause 4.17.5	Clause 4.17.5
Clause 4.20.7	Clause 4.20.7	Clause 4.20.7	Clause 4.17.6	Clause 4.17.6
Clause 4.20.8	Clause 4.20.8	Clause 4.20.8	Clause 4.17.7	Clause 4.17.7
Clause 4.21	Clause 4.21	Clause 4.21	Clause 4.18	Clause 4.18
Clause 4.21.1	Clause 4.21.1	Clause 4.21.1	Clause 4.18.1	Clause 4.18.1
Clause 4.22	Clause 4.22	Clause 4.22	Clause 4.19	Clause 4.19

10 Variations

10.1 Introduction

Variations under the sub-contract are changes to the scope of the work or conditions under which the work is carried out, only to the extent that terms in the sub-contract provide for such changes.

This is not the same as variations to the sub-contract. Variations to the sub-contract are changes to the sub-contract agreement and/or its terms and conditions, and such changes can only be made by express agreement between the parties.

The nature of construction works makes the possibility of changes to the scope of the works or the conditions under which the work is carried out susceptible to change. It is therefore necessary that the sub-contract conditions incorporate provisions for variations to the original scope of the work or the conditions under which the work is to be carried out.

If such a provision were not included within the sub-contract, then every time a change occurred, this would constitute a breach of the sub-contract or would require separate sub-contracts to be raised for all additional works and other changes.

Against this background, in this chapter, the following matters will be dealt with, using the sub-contract clause references, etc. contained in the SBCSub/D/A or SBCSub/D/C (as appropriate). Whilst it is clearly beyond the scope of this book to review every nuance of the other sub-contract forms under consideration, the equivalent provisions (where applicable) within the SBCSub/A or SBCSub/C, the DBSub/A or DBSub/C, the ICSub/A or ICSub/C and the ICSub/D/A or ICSub/D/C are given at the table at the end of this chapter. It must be emphasised that before considering a particularly issue, the actual terms of the appropriate edition of the relevant sub-contract should be reviewed by the reader (and/or legal advice should be sought as appropriate) before proceeding with any action/inaction in respect of the sub-contract in question.

In this chapter the following matters are dealt with:

- What is a variation?
- How is a variation instructed?
- Can a variation vitiate a contract?

The JCT 2011 Building Sub-contracts, First Edition. Peter Barnes and Matthew Davies.
© 2016 John Wiley & Sons, Ltd. Published 2016 by John Wiley & Sons, Ltd.

- Must a sub-contractor comply with all variation directions issued?
- Must a sub-contractor comply with all variation directions issued that may injuriously affect the efficacy of the sub-contractor's design?
- What happens if a sub-contractor does not comply with a direction issued?
- How should variations be valued?
- How should variations that relate to a sub-contractor's design work be valued?
- What is the procedure to be followed in respect of a schedule 2 quotation?
- What are the valuation rules?

10.2   What is a variation?

The definitions section (clause 1.1) defines a variation as 'see clause 5.1', and under clause 5.1, a variation is defined as being:

'5.1.1 The alteration or modification of the design, the quality or (except where the Remeasurement basis applies) the quantity of the Sub-contract Works including:

1. the addition, omission or substitution of any work;
2. the alteration of the kind or standard of any of the materials or goods to be used in the Sub-contract Works;
3. the removal from the site of any work executed or materials or goods brought thereon by the Sub-contractor for the purposes of the Sub-contract Works other than work, materials or goods which are not in accordance with this Sub-contract.

5.1.2 The imposition in an instruction of the Architect/Contract Administrator issued under the Main Contract (or a direction of the Contractor passing on that instruction) of any obligation or restrictions in regard to the matters set out in this clause 5.1.2 or the addition to or alteration or omission of any such obligations or restrictions so imposed or imposed by the Numbered Documents and the Schedule of Information and its annexures in regard to:

1. access to the site or use of any specific parts of the site;
2. limitations of working space;
3. limitations of working hours;
4. the execution or completion of the work in any specific order'.

10.3   How is a variation instructed?

All variation instructions are expected to be issued in writing in line with clause 3.4.

If the contractor purports to give any direction to the sub-contractor, or his or her authorised representative, other than in writing (e.g. orally), then clause 3.7 makes it clear that that direction shall have no immediate effect.

However, a procedure is set out under clause 3.7 that can be followed for the confirmation of instructions issued other than in writing.[1]

This procedure is:

(1)   The sub-contractor is to confirm the direction to the contractor in writing, within 7 days of the direction being issued.
(2)   If the contractor does not dissent in writing to the sub-contractor that the direction was issued (within 7 days from receipt of the sub-contractor's confirmation), then the direction shall take effect as from the expiry of that said 7-day period.

If, within 7 days of giving a direction other than in writing, the contractor confirms the direction in writing, then the sub-contractor shall not be obliged to confirm the direction and the direction shall take effect as from the date of the contractor's written confirmation (as clause 3.7.1).

It should be noted that if action is taken on the basis of an oral instruction (more commonly referred to as a verbal instruction), without following the above procedure, then such action is taken entirely at the sub-contractor's risk.

Notwithstanding the above, clause 3.7.2 confirms that if a sub-contractor complies with an oral direction without following the procedure set out under clause 3.7, the contractor may at any time prior to the final payment under the sub-contract confirm the direction in writing with retrospective effect (refer to clause 3.7.2). The operative word in the foregoing phrase is 'may', and there is no obligation for a contractor to retrospectively confirm a purported direction at all, and in reality, it is extremely unlikely that a contractor would confirm a purported direction where he or she disagrees that such a direction was issued in the first place.

It should be noted that any written instruction of the architect/contract administrator issued under the main contract that affects the sub-contract works which is then issued by the contractor to the sub-contractor shall be deemed to be a direction of the contractor (refer to clause 3.4). How this latter requirement will be implemented in practice, in terms of disclosure by the contractor, remains to be seen.

It should be noted that certain directions issued by the contractor under clauses 6.7.3, 6.8.3.1 and 6.14.1 (in respect of particular insurance-related matters) are to be treated as variations in line with clauses 6.7.4, 6.8.3.2 and 6.14.2, respectively.

10.4   Can a variation vitiate a contract?

To vitiate means to make invalid or ineffectual. It is quite common for contracts to contain a clause which expressly states that no variation shall vitiate the contract. Although the above phrase is common in most standard forms,

---

[1]No oral/verbal instructions are allowed under the ICSub or the ICSub/D.

it is submitted that there must be some limit to the nature and extent of a variation which can be ordered, particularly where the variation requires the sub-contractor to carry out design works or works on site for which he or she professes no skill or experience whatsoever.

10.5   Must a sub-contractor comply with all variation directions issued?

Clause 3.4 expressly provides that no variation directed by the contractor or subsequently sanctioned by him or her shall vitiate the sub-contract, and clause 3.5 notes that the sub-contractor is to comply with all directions issued to him or her under clause 3.4.

Clause 3.5.1 notes that where a contractor issues a direction which requires a variation of the type referred to in clause 5.1.2 (i.e. the imposition, alteration or omission of any obligation or restriction in regard to access to the site, limitation of working space, limitation of working hour), the sub-contractor need not comply to the extent that he or she makes reasonable objection to it in writing to the contractor.

Of course, there could be many disputes regarding what 'reasonable objections' are, and any such dispute could have a major impact on the actions that a sub-contractor can or should sensibly attempt to take.

In the case where a direction is given which pursuant to clause 5.3.1 requires the sub-contractor to provide a variation quotation (a schedule 2 quotation) the variation shall not be carried out until the contractor has in relation to the schedule 2 quotation either issued written acceptance or has issued a further direction under clause 5.3.2 (i.e. that the work is to be valued by a valuation by the contractor in accordance with the valuation rules pursuant to clause 5.2).

10.6   Must a sub-contractor comply with all variation directions issued that may injuriously affect the efficacy of the sub-contractor's design?

This issue only arises in the sub-contracts that contain an element of design.

Thus, under clause 3.5.3, if, in the sub-contractor's opinion, compliance with any direction of the contractor may injuriously affect the efficacy of the design of the sub-contractor's design, the sub-contractor has the right to object (provided that such objection is made within 5 days from receipt of the contractor's direction). The contractor's direction will then not have any effect unless confirmed by the contractor.

Naturally, given the strict liability placed upon the sub-contractor in respect of design, the sub-contractor would need to make clear to the contractor that, by confirming the earlier direction that was (in the opinion of the sub-contractor) injurious to the sub-contractor's design, the contractor would be removing the liability for such design from the sub-contractor.

10.7   What happens if a sub-contractor does not comply with a direction issued?

Clause 3.6 makes it clear that if, within 7 days after receipt of a written notice from the contractor which requires a sub-contractor to comply with a direction, the sub-contractor does not comply with that direction, then the contractor may employ and pay other persons to execute any work whatsoever which may be necessary to give effect to that direction.

In such a situation, the sub-contractor would be liable for all additional costs incurred by the contractor in connection with such employment and an appropriate deduction shall either be taken into account in the calculation of the final sub-contract sum or shall be recoverable by the contractor from the sub-contractor as a debt.

10.8   How should variations be valued?

Unless otherwise agreed by the contractor and the sub-contractor, clause 5.2 states that variations are valued either by way of an accepted schedule 2 quotation or in accordance with clauses 5.6–5.12 (known as the Valuation Rules).

If the sub-contract is based on the adjustment basis, then in line with clause 5.2.1 of the conditions, the valuation rules will apply to:

- all variations, including any sanctioned in writing by the contractor but excluding any variation to an accepted schedule 2 quotation;
- all work which under the conditions is to be treated as a variation;
- all work executed by the sub-contractor in accordance with the directions of the contractor as to the expenditure of Provisional Sums which are included in the sub-contract documents;
- all work executed by the sub-contractor for which an approximate quantity has been included in any bills of quantities or in the contract requirements.

If the sub-contract is based on the re-measurement basis, then in line with clause 5.2.1 of the conditions, the valuation rules will apply to all work executed by the sub-contractor in accordance with the sub-contract documents and the directions of the contractor, including any direction requiring a variation or in regard to the expenditure of a Provisional Sum included in the sub-contract documents.

Effect to the valuation of variations shall be given in the calculation of the final sub-contract sum (clause 5.5).

10.9   How should variations that relate to a sub-contractor's design work be valued?

This issue only arises in the sub-contracts that contain an element of design.

Thus, under clause 5.10, such valuation shall be in accordance with that clause, and references in clauses 5.6 and 5.7 shall exclude the valuation of variations in respect of the sub-contractor's designed works.

Valuations under clause 5.10 are to be based on the following:

- Allowance shall be made in such valuation for work involved in the preparation of the relevant design work.
- The valuation shall be consistent with the values of work of a similar character set out in the sub-contractor's designed works analysis, making due allowance for any change in the conditions under which work is carried out and/or any significant change in the quantity of work so set out.

Where there is no work of a similar character set out in the sub-contractor's designed works analysis, a fair valuation shall be made (clause 5.10.2):

- The valuation of the omission of the work set out in the sub-contractor's designed works analysis in accordance with the values therein for such work (clause 5.10.3).
- The Variation general rules, daywork provisions and the change of condition of other work affected by a variation shall apply so far as is relevant.

10.10 What is the procedure to be followed in respect of a schedule 2 quotation?

A schedule 2 quotation is a quotation that is provided in line with the provisions of schedule 2 of the sub-contract.

Clause 5.3.1 notes that if the contractor in his or her direction states that the sub-contractor is to provide a schedule 2 quotation, the sub-contractor shall, subject to receipt of sufficient information, provide a quotation, unless, within 4 days of his or her receipt of that direction (or such longer period as is either stated in the direction or is agreed between the contractor and the sub-contractor), he or she notifies the contractor that he or she disagrees with the application of that procedure to that direction.

If the sub-contractor does issue such a notice of disagreement, then the sub-contractor shall not be obliged to provide a schedule 2 quotation, and the work shall not be carried out unless and until the contractor gives a further direction that the work is to be carried out and is to be valued using one of the other valuation rules.

10.10.1 Submission of quotation

In summary, and assuming that the contractor and the sub-contractor do not agree upon extended time periods, then where schedule 2 quotations are required:

- the contractor in his or her direction is to state that the sub-contractor is to provide a schedule 2 quotation.
- if the sub-contractor disagrees with the application of that procedure to that direction, he or she is to notify the contractor of that fact within 4 days of his or her receipt of the direction.

- if the sub-contractor does issue such a notice of disagreement, then the sub-contractor shall not be obliged to provide a schedule 2 quotation, and the work shall not be carried out unless and until the contractor gives a further direction that the work is to be carried out and is to be valued using one of the valuation rules.
- if the sub-contractor does not issue such a notice of disagreement, but considers that the information provided is not sufficient, then, not later than 4 days from the date of the instruction, he or she shall notify the contractor, who shall supply the necessary information.
- the sub-contractor shall submit his or her schedule 2 quotation to the contractor not later than 14 days from the later of:
  - ○ the date of receipt of the instruction; or
  - ○ the date of receipt by the sub-contractor of sufficient information as referred to above.
- the schedule 2 quotation shall remain open for acceptance by the contractor for 14 days from its receipt by the contractor.

10.10.2   Content of the quotation

It is important to note that the schedule 2 quotation is to separately comprise:

- the amount of the adjustment to the final sub-contract sum (excluding any amount to be paid in lieu of any ascertainment for direct loss and/or expense, but including allowance, where appropriate, for preliminary items) supported by all necessary calculations, which shall be made by reference, where relevant, to the rates and prices in the sub-contract sum or the sub-contract tender sum.
- any adjustment to the time required for completion of the sub-contract works and/or of any works in any Section by reference to the period or periods stated in the sub-contract particulars (item 5) to the extent that such adjustment is not included in any revision to the completion date previously issued by the contractor or in the contractor's confirmed acceptance of any other schedule 2 quotation.
- the amount to be paid in lieu of any ascertainment, under clause 4.19 of direct loss and expense not included in any other accepted schedule 2 quotation or in any previous ascertainment under clause 4.19.
- a fair and reasonable amount in respect of the cost of preparing the schedule 2 quotation.
- where specifically required by the instruction, the sub-contractor shall provide indicative information in statements on:
  - (a)   the additional resources (if any) required to carry out the variation;
  - (b)   the method of carrying out the variation.

Each part of the schedule 2 quotation shall contain reasonably sufficient supporting information to enable that part to be evaluated by or on behalf of the contractor.

10.10.3 Acceptance of the quotation

If the contractor wishes to accept a schedule 2 quotation, the contractor is to notify the sub-contractor in writing not later than the last day of the period of acceptance.

If the contractor accepts a schedule 2 quotation, the contractor shall, immediately upon that acceptance, confirm such acceptance in writing to the sub-contractor:

(1) that the sub-contractor is to carry out the variation;
(2) the adjustment to the final sub-contract sum;
(3) any adjustment to the time required by the sub-contractor for completion of the sub-contract works or (where applicable) such works in any relevant section.

10.10.4 Quotation not accepted

If the contractor does not accept the schedule 2 quotation by the expiry of the period of acceptance, the contractor shall, on the expiry of that period, either:

(1) instruct that the variation is to be carried out and is to be valued under the valuation rules; or
(2) instruct that the variation is not to be carried out.

10.10.5 Cost of quotation

If a schedule 2 quotation is not accepted, a fair and reasonable amount shall be added to the final sub-contract sum in respect of the preparation of the schedule 2 quotation provided that the schedule 2 quotation has been prepared on a fair and reasonable basis.

Although there is no definition of when a schedule 2 quotation has been prepared on a fair and reasonable basis, it is clear that the non-acceptance by the contactor shall not, of itself, be evidence that the quotation was not prepared on a fair and reasonable basis.

10.10.6 Restriction on use of quotation

Unless the contractor accepts a schedule 2 quotation, neither the contractor nor the sub-contractor may use that quotation for any purpose whatsoever.

10.10.7 What happens if work covered by a schedule 2 quotation that has been accepted by the contractor is itself varied?

Where a schedule 2 quotation has been accepted by the contractor, then if the contractor subsequently issues a direction varying the work that was the subject matter of that schedule 2 quotation, the contractor shall make a valuation of that variation on a fair and reasonable basis having regard to the content of the schedule 2 quotation and shall include in that valuation the direct loss and/or expense, if any, incurred by the sub-contractor because the regular progress of the sub-contract works or any Section thereof is materially affected by compliance with the direction (refer to clause 5.3.3).

10.11   What are the valuation rules?

The valuation rules are applicable to the extent that a valuation relates to the execution of additional or substituted work which can properly be valued by measurement.

It should be noted that in respect of all the valuation rules, the contractor is to give the sub-contractor an opportunity to be present and to allow the sub-contractor to take such notes and measurements as the sub-contractor may require at the time when it is necessary to measure work for the purpose of valuation (refer to clause 5.4).

There are certain general rules that apply, as clause 5.8, namely:

(1)   Where there are bills of quantities, the measurement of variations shall be in accordance with the same principles as those governing the preparation of those bills of quantities.

The default position in respect of the preparation of bills of quantities is that *the Standard Method of Measurement of Building Works, 7th Edition* (SMM7), produced by the Royal Institution of Chartered Surveyors and the Construction Confederation, will have been used.

Clause 2.9.1 notes that if there is any unstated departure from SMM7 in the preparation of the bills of quantities, or if there is any error in description or in quantity or any omission of items, the departure, error or omission shall be corrected, and any such correction shall be treated as a variation.

(2)   When valuing variations, allowances shall be made for any percentage or lump sum adjustments in any bills of quantities and/or other sub-contract documents.

(3)   Where the adjustment basis applies, an allowance (where appropriate) shall be made for any addition to or reduction from any preliminary items of the type referred to in SMM7. However, the above principle does not apply where a variation relates to the contractor's direction for the expenditure of a Provisional Sum for defined work, for the reasons outlined below.

The SMM7 General Rule 10.3 defines a Provisional Sum for defined work as being a sum for work which is not completely designed but for which the following information is provided:

(a)   The nature of the construction of the work

(b)   A statement of how and where the work is fixed to the building and what other work is to be fixed thereto

(c)   A quantity or quantities which indicate the scope and extent of the works

(d) Any specific limitations on the method or the sequence or the timing of the works

If the information specified in rule 10.3 is not available, the contract bills should describe the Provisional Sum as undefined.

General Rule 10.4 of SMM7 states that: 'where Provisional Sums are given for defined work the Contractor (and the sub-contractor) will be deemed to have made due allowance in programming, planning and pricing preliminaries', whilst General Rule 10.6 of SMM7 states that the contractor is deemed to have made *no such allowance* for undefined work. Therefore, the categorisation of defined or undefined Provisional Sums has significant implications.

If a Provisional Sum is incorrectly described as defined, then clause 2.9.1 makes it clear that if there is any error in or omission of information in any item which is the subject of a Provisional Sum for defined work, the description shall be corrected so that it does provide that information. Clause 2.9.3 then adds that any such correction, alteration or modification to the Provisional Sum shall be treated as a variation.

(4) Where the re-measurement basis applies, any amounts priced in the preliminaries section of any bills of quantities shall be adjusted (where appropriate) to take into account any variations of any contractor's directions for the expenditure of Provisional Sums for undefined work included in the sub-contract documents.

10.11.1 The valuation rule dealt with under clauses 5.6.1.1 and 5.7.1 – variation works of similar character, similar conditions and similar quantity

Under this valuation rule, where variation work is carried out that is of similar character to, is executed under similar conditions as and does not significantly change the quantity of work set out in any bills of quantities and/or other sub-contract documents, the rates and prices for the work in those documents shall determine the valuation. This is subject only to clause 5.10 in the case of sub-contractor design works.

Similar conditions are those conditions which are to be derived from the express provisions of the sub-contract. Extrinsic evidence of, for instance, the parties' subjective expectations is not admissible.[2]

10.11.2 The valuation rule dealt with under clauses 5.6.1.2 and 5.7.2 – variation works of similar character but dissimilar conditions and/or dissimilar quantity

Under this valuation rule, where variation work is carried out that is of similar character to work set out in any bills of quantities and/or other sub-contract

---

[2] *Wates Construction v. Bredero Fleet* (1993) 63 BLR 128.

documents, but is not executed under similar conditions thereto and/or significantly changes its quantity, the rates and prices for the work set out in the documents above shall be the basis for determining the valuation, and the valuation value shall include a fair allowance for such difference in conditions and/or quantity.

It must be noted that where a sub-contractor has simply submitted a rate in error, it has been found that such a mistake would not prevent the use of those rates to value a subsequent variation.[3]

10.11.3   The valuation rule dealt with under clauses 5.6.1.3 and 5.7.3 – variation works not of similar character

Under this valuation rule, where variation work is carried out that is not of similar character to work set out in any bills of quantities and/or other sub-contract documents, the work shall be valued at fair rates and prices.

Fair rates and prices was considered in the *Crittall Windows* v. *TJ Evers*[4] case. In that case, Judge Humphrey Lloyd QC stated that 'a fair valuation generally means a valuation which will not give the contractor more than his or her actual costs reasonably and necessarily incurred plus similar allowance for overheads and profit'. Judge Humphrey Lloyd QC has since repeated this approach in two other cases.[5]

However, a more common view is that fair rates and prices must have regard to the contractor's general pricing level, and therefore, a valuation below actual cost would be fair where the contract price is below actual costs or market pricing.

10.11.4   The valuation rule dealt with under clause 5.6.1.4 – reasonably accurate approximate quantity

This valuation rule is only applicable when the sub-contract is let on the adjustment basis (i.e. where a sub-contract sum is used as the basis of the sub-contract rather than where a sub-contract tender sum is used). This rule is that where an approximate quantity is included in the bills of quantities and/or other sub-contract documents, and that approximate quantity is a reasonably accurate forecast of the quantity of work actually required, then the rate or price for that approximate quantity shall determine the valuation of the varied works.

This rule only applies where the work required has not been altered or modified in any way from that specified other than in terms of quantity.

10.11.5   The valuation rule dealt with under clause 5.6.1.5 – not reasonably accurate approximate quantity

This valuation rule is again also only applicable when the sub-contract is let on the adjustment basis. This rule is that where an approximate quantity is

---

[3]*Henry Boot Construction Ltd* v. *Alstom* [1999] BLR 123.

[4]*Crittall Windows* v. *TJ Evers Ltd* (1996) 54 Con LR 66.

[5]*Floods Queensferry Ltd* v. *Shand Construction Ltd* [1999] BLR 315; *Weldon Plant Ltd* v. *The Commissioner for the New Towns* [2000] BLR 496.

included in the bills of quantities and/or other sub-contract documents, and that approximate quantity is not a reasonably accurate forecast of the quantity of work actually required, then the rate or price for that approximate quantity shall be used as the basis for determining the valuation of the varied works, with a fair allowance being made for such difference in quantity.

Again, this rule only applies where the work required has not been altered or modified in any way from that specified other than in terms of quantity.

10.11.6   The valuation rule dealt with under clause 5.6.2 – omission of work

This valuation rule is only applicable when the sub-contract is let on the adjustment basis. This rule states that to the extent that a valuation relates to the omission of work set out in any bills of quantities and/or other sub-contract documents, the rates and the prices for such work set out therein (and no adjustment to those rates) shall determine the valuation of the work omitted. This is subject only to clause 5.10, in the case of sub-contractor design works.

10.11.7   Change of conditions for other work

This valuation rule (under clause 5.11) notes that if there is a substantial change in the conditions under which any other work (including sub-contractor design works) is executed, as a result of:

(1)   compliance with any direction requiring a variation (except a variation for which a schedule 2 quotation has been accepted or where the variation is to works covered by a schedule 2 quotation; and except for any direction as a result of which work due to be executed in the sub-contract on the re-measurement basis is not executed);

(2)   compliance with any direction as to the expenditure of a Provisional Sum for undefined work;

(3)   compliance with any direction as to the expenditure of a Provisional Sum for defined work, to the extent that the direction for that work differs from the description given for such work in any bills of quantities; or

(4)   where the adjustment basis applies, the execution of the work for which an approximate quantity is included in any bills of quantities, to the extent that the quantity is more or less than the quantity ascribed to that work in those bills; then such other work shall be treated as if it had been the subject of a direction of the contractor requiring a variation and shall be valued on the basis of the rates and prices for the work set out in any bills of quantities and/or other sub-contract documents, together with a fair allowance for the difference in conditions.

10.11.8   Daywork

'Daywork' is a means of valuing variations which is based upon recorded time, and recorded material and plant usage. Clause 5.9 makes it clear that Daywork should only be used when a variation cannot properly be valued by measurement.

The time and resources spent on a variation that it is proposed is valued on a Daywork basis should be recorded by the sub-contractor on a Daywork voucher (more normally referred to as a Daywork sheet), and that Daywork voucher should have sections to record the operatives' names and trades, the time spent by the operatives on the variation, and also the plant and materials used.

The completed Daywork voucher must be delivered for verification to the contractor not later than five business days after it was executed (clause 5.9). The reason for this relatively short timetable is clearly to enable the contractor to contemporaneously check the details on the Daywork voucher.

The various elements of the Daywork voucher are valued on the basis of the schedule of Daywork rates included in the numbered documents. If no such schedule is included in the numbered documents, then the various elements of the Daywork voucher are valued on the following basis:

- Labour

    Either at the all-in rate stated for each grade of operative as noted under item 12 of the sub-contract particulars or, if an all-in rate is not stated, then the prime cost for each grade of operative calculated in accordance with the Definition of Prime Cost of Daywork carried out under a Building Contract, together with percentage additions to the prime cost at the rates set out in the sub-contract particulars item 12 (as clause 5.9.1)

    The Definition of Prime Cost of Daywork carried out under a Building Contract is a document issued by the Royal Institution of Chartered Surveyors and the Construction Confederation, with the edition to be used being that current at the sub-contract base date (as stated in the sub-contract particulars).

    If the operatives in question are within the province of any specialist trade where the Royal Institution of Chartered Surveyors and the appropriate body representing the employers in that trade have agreed and issued a definition of prime cost of Daywork, then the prime cost for each grade of operative calculated in accordance with that definition is to be used, together with the percentage addition to the prime cost at the rates set out in the sub-contract particulars (as clause 5.9.2).

    There are currently three definitions to which clause 5.9.2 refers, being those agreed between the Royal Institution of Chartered Surveyors and the Electrical Contractors Association, the Electrical Contractors Association of Scotland, and the Heating and Ventilating Association, respectively.

- Plant

    Plant is valued in accordance with one or more of the three definitions noted above to obtain the plant Daywork base rate at the time that the Daywork was carried out, and then the appropriate percentage addition is applied as inserted against the appropriate definition (as clause 5.9.1).

    When tendering for work, it is important to note that the prime cost for plant on a Daywork basis may, on occasion, be based on a basic plant schedule which may pre-date the date when the Dayworks are being carried out by several years, and the percentage addition allowed by the sub-contractor needs to reflect this fact.

■ Materials

Materials are priced at the prime cost for materials, together with the percentage addition to the prime cost at the rates set out in the sub-contract particulars item 12 (as clause 5.9.1).

10.11.9   Fair valuation

If a valuation does not relate to the execution of work or the omission of work, or if the valuation of any work or liabilities directly associated with a variation cannot reasonably be effected in the valuation by any of the valuation rules noted above, then clause 5.12.1 notes that a fair valuation shall be made.

It must be noted that this additional provision does not relate to the execution of additional work or the omission of work, and only relates to any work or liabilities directly associated with a variation.

10.11.10   Non-recovery of loss and/or expense

Clause 5.12.2 notes that no allowance shall be made under the valuation rules for any effect upon the regular progress of the works (or any part of them) or for any other direct loss and/or expense for which the sub-contractor would be reimbursed by any other provision in the sub-contract.

10.12   Equivalent sub-contract provisions

In the table below, the equivalent sub-contract provisions to that within the SBCSub/D/A or SBCSub/D/C in the text above are listed for the SBCSub/A or SBCSub/C, DBSub/A or DBSub/C, ICSub/A or ICSub/C and ICSub/D/A or ICSub/D/C.

SBCSub/D/A or SBCSub/D/C	SBCSub/A or SBCSub/C	DBSub/A or DBSub/C	ICSub/A or ICSub/C	ICSub/D/A or ICSub/D/C
Sub-contract Particulars Item 5	Sub-contract Particulars Item 5	Sub-contract Particulars Item 5	Sub-contract Particulars Item 5	Sub-contract Particulars Item 5
Sub-contract Particulars Item 12	Sub-contract Particulars Item 12	Sub-contract Particulars Item 12	Sub-contract Particulars Item 11	Sub-contract Particulars Item 11
Schedule 2 Quotation	Schedule 2 Quotation	Schedule 2 Quotation	Not applicable to this Sub-contract as Schedule 2 Quotations are not provided for under this Sub-contract	Not applicable to this Sub-contract as Schedule 2 Quotations are not provided for under this Sub-contract

SBCSub/D/A or SBCSub/D/C	SBCSub/A or SBCSub/C	DBSub/A or DBSub/C	ICSub/A or ICSub/C	ICSub/D/A or ICSub/D/C
Clause 1.1	Clause 1.1	Clause 1.1	Clause 1.1	Clause 1.1
Clause 2.9.1	Clause 2.9.1	Clause 2.9.1	Clause 2.7.2	Clause 2.7.2
Clause 2.9.3	Clause 2.9.2	Clause 2.9.3	Clause 2.9	Clause 2.9
Clause 3.4	Clause 3.4	Clause 3.4	Clause 3.4	Clause 3.4
Clause 3.5	Clause 3.5	Clause 3.5	Clause 3.5	Clause 3.5
Clause 3.5.1	Clause 3.5.1	Clause 3.5.1	Clause 3.5	Clause 3.5
Clause 3.5.3	Clause 3.5.3	Clause 3.5.3	Clause 3.5	Clause 3.5
Clause 3.6	Clause 3.6	Clause 3.6	Clause 3.6	Clause 3.6
Clause 3.7	Clause 3.7	Clause 3.7	Not applicable to this Sub-contract as oral instructions are not provided for under this Sub-contract	Not applicable to this Sub-contract as oral instructions are not provided for under this Sub-contract
Clause 3.7.1	Clause 3.7.1	Clause 3.7.1	Not applicable to this Sub-contract as oral instructions are not provided for under this Sub-contract	Not applicable to this Sub-contract as oral instructions are not provided for under this Sub-contract
Clause 3.7.2	Clause 3.7.2	Clause 3.7.2	Not applicable to this Sub-contract as oral instructions are not provided for under this Sub-contract	Not applicable to this Sub-contract as oral instructions are not provided for under this Sub-contract
Clause 4.19	Clause 4.19	Clause 4.19	Clause 4.16	Clause 4.16
Clause 5.1	Clause 5.1	Clause 5.1	Clause 5.1	Clause 5.1
Clause 5.1.2	Clause 5.1.2	Clause 5.1.2	Clause 5.1.2	Clause 5.1.2
Clause 5.2	Clause 5.2	Clause 5.2	Clause 5.2	Clause 5.2
Clause 5.2.1	Clause 5.2.1	Clause 5.2.1	Clause 5.2.1	Clause 5.2.1
Clause 5.3.1	Clause 5.3.1	Clause 5.3.1	Not applicable to this Sub-contract as Schedule 2 Quotations are not provided for under this Sub-contract	Not applicable to this Sub-contract as Schedule 2 Quotations are not provided for under this Sub-contract
Clause 5.3.2	Clause 5.3.2	Clause 5.3.2	Not applicable to this Sub-contract as Schedule 2 Quotations are not provided for under this Sub-contract	Not applicable to this Sub-contract as Schedule 2 Quotations are not provided for under this Sub-contract

(Continued)

SBCSub/D/A or SBCSub/D/C	SBCSub/A or SBCSub/C	DBSub/A or DBSub/C	ICSub/A or ICSub/C	ICSub/D/A or ICSub/D/C
Clause 5.3.3	Clause 5.3.3	Clause 5.3.3	Not applicable to this Sub-contract as Schedule 2 Quotations are not provided for under this Sub-contract	Not applicable to this Sub-contract as Schedule 2 Quotations are not provided for under this Sub-contract
Clause 5.4	Clause 5.4	Clause 5.4	Not applicable to this Sub-contract as the Sub-contractor is not specifically to be given the opportunity to be present to make notes and/or carry out measurements in respect of Variations under this Sub-contract	Not applicable to this Sub-contract as the Sub-contractor is not specifically to be given the opportunity to be present to make notes and/or carry out measurements in respect of Variations under this Sub-contract
Clause 5.5	Clause 5.5	Clause 5.5	Not applicable to this Sub-contract because it in part relates to Schedule 2 Quotations, and Schedule 2 Quotations are not provided for under this Sub-contract	Not applicable to this Sub-contract because it in part relates to Schedule 2 Quotations, and Schedule 2 Quotations are not provided for under this Sub-contract
Clause 5.6	Clause 5.6	Clause 5.6	Clause 5.3	Clause 5.3
Clause 5.6.1.1	Clause 5.6.1.1	Clause 5.6.1.1	Clause 5.3.1.1	Clause 5.3.1.1
Clause 5.6.1.2	Clause 5.6.1.2	Clause 5.6.1.2	Clause 5.3.1.2	Clause 5.3.1.2
Clause 5.6.1.3	Clause 5.6.1.3	Clause 5.6.1.3	Clause 5.3.1.3	Clause 5.3.1.3
Clause 5.6.1.4	Clause 5.6.1.4	Clause 5.6.1.4	Clause 5.3.1.4	Clause 5.3.1.4
Clause 5.6.1.5	Clause 5.6.1.5	Clause 5.6.1.5	Clause 5.3.1.5	Clause 5.3.1.5
Clause 5.6.2	Clause 5.6.2	Clause 5.6.2	Clause 5.3.2	Clause 5.3.2
Clause 5.7	Clause 5.7	Clause 5.7	Clause 5.4	Clause 5.4
Clause 5.7.1	Clause 5.7.1	Clause 5.7.1	Clause 5.4.1	Clause 5.4.1
Clause 5.7.2	Clause 5.7.2	Clause 5.7.2	Clause 5.4.2	Clause 5.4.2
Clause 5.7.3	Clause 5.7.3	Clause 5.7.3	Clause 5.4.3	Clause 5.4.3
Clause 5.8	Clause 5.8	Clause 5.8	Clause 5.5	Clause 5.5
Clause 5.9	Clause 5.9	Clause 5.9	Clause 5.6	Clause 5.6
Clause 5.9.1	Clause 5.9.1	Clause 5.9.1	Clause 5.6.1	Clause 5.6.1
Clause 5.9.2	Clause 5.9.2	Clause 5.9.2	Clause 5.6.2	Clause 5.6.2

SBCSub/D/A or SBCSub/D/C	SBCSub/A or SBCSub/C	DBSub/A or DBSub/C	ICSub/A or ICSub/C	ICSub/D/A or ICSub/D/C
Clause 5.10	Not applicable to this Sub-contract as it relates to the valuation of variations to the design work carried out by the Sub-contractor, and Sub-contract design is not provided for under this Sub-contract	Clause 5.10	Not applicable to this Sub-contract as it relates to the valuation of variations to the design work carried out by the Sub-contractor, and Sub-contract design is not provided for under this Sub-contract	Clause 5.9
Clause 5.10.2	Not applicable to this Sub-contract as it relates to the valuation of variations to the design work carried out by the Sub-contractor, and Sub-contract design is not provided for under this Sub-contract	Clause 5.10.2	Not applicable to this Sub-contract as it relates to the valuation of variations to the design work carried out by the Sub-contractor, and Sub-contract design is not provided for under this Sub-contract	Clause 5.9.2
Clause 5.10.3	Not applicable to this Sub-contract as it relates to the valuation of variations to the design work carried out by the Sub-contractor, and Sub-contract design is not provided for under this Sub-contract	Clause 5.10.3	Not applicable to this Sub-contract as it relates to the valuation of variations to the design work carried out by the Sub-contractor, and Sub-contract design is not provided for under this Sub-contract	Clause 5.9.3
Clause 5.11	Clause 5.11	Clause 5.11	Clause 5.7	Clause 5.7
Clause 5.12	Clause 5.12	Clause 5.12	Clause 5.8	Clause 5.8
Clause 5.12.1	Clause 5.12.1	Clause 5.12.1	Clause 5.8.1	Clause 5.8.1
Clause 5.12.2	Clause 5.12.2	Clause 5.12.2	Clause 5.8.2	Clause 5.8.2
Clause 6.7.3	Clause 6.7.3	Clause 6.7.3	Clause 6.7.3	Clause 6.7.3
Clause 6.7.4	Clause 6.7.4	Clause 6.7.4	Clause 6.7.4	Clause 6.7.4
Clause 6.8.3.1	Clause 6.8.3.1	Clause 6.8.3.1	Clause 6.8.3.1	Clause 6.8.3.1
Clause 6.8.3.2	Clause 6.8.3.2	Clause 6.8.3.2	Clause 6.8.3.2	Clause 6.8.3.2
Clause 6.14.1	Clause 6.14.1	Clause 6.14.1	Clause 6.14.1	Clause 6.14.1
Clause 6.14.2	Clause 6.14.2	Clause 6.14.2	Clause 6.14.2	Clause 6.14.2

11 Injury, Damage and Insurance

11.1 Introduction

Obviously, one of the major issues on construction projects is the risk involved with injury and/or damage.

It is important that the risk of injury or damage is retained by the party best able to deal with that risk, and it is obviously equally important that the risks are properly and adequately insured where possible.

Against this background, in this chapter, the following matters will be dealt with, using the sub-contract clause references, etc. contained in the SBCSub/D/A or SBCSub/D/C (as appropriate). Whilst it is clearly beyond the scope of this book to review every nuance of the other sub-contract forms under consideration, the equivalent provisions (where applicable) within the SBCSub/A or SBCSub/C, the DBSub/A or DBSub/C, the ICSub/A or ICSub/C and the ICSub/D/A or ICSub/D/C are given at the table at the end of this chapter. It must be emphasised that before considering a particularly issue, the actual terms of the appropriate edition of the relevant sub-contract should be reviewed by the reader (and/or legal advice should be sought as appropriate) before proceeding with any action/inaction in respect of the sub-contract in question.

In this chapter we will consider how the sub-contract deals with:

- The Sub-contractor's liability for personal injury or death.
- The Sub-contractor's liability for injury or damage to property.
- The Sub-contractor's liability for loss or damage to the Sub-contract works.
- What Specified Perils cover, for loss or damage to works and site materials, does the Sub-contractor obtain under the joint names all risks policies?
- What is the Sub-contractor's liability for damage to the Sub-contract works?
- What are the employer's options where terrorism cover is not available?
- Is the contractor responsible for damage caused to Sub-contractor's plant etc.?
- Is the Sub-contractor required to take out professional indemnity insurance?
- When and how does the Joint Fire Code apply?

The JCT 2011 Building Sub-contracts, First Edition. Peter Barnes and Matthew Davies.
© 2016 John Wiley & Sons, Ltd. Published 2016 by John Wiley & Sons, Ltd.

11.2   Sub-contractor's liability for personal injury or death

11.2.1   Indemnity

The Sub-contractor is liable for, and shall indemnify the contractor against any expense, liability, loss, claim or proceedings whatsoever in respect of personal injury to or the death of any person arising out of, or in the course of or caused by the carrying out of the Sub-contract works (as clause 6.2 of the Sub-contract conditions). However, clause 6.5.2 notes that this indemnity does not include for personal injury to or the death of any person due to the effect of an Excepted Risk as defined under clause 6.1 of the sub-contract conditions.

Clause 6.2 of the Sub-contract conditions states that the Sub-contractor is not liable where the personal injury to or the death of any person arising out of, or in the course of or caused by the carrying out of the Sub-contract works is due to any act of neglect, breach of statutory duty, omission or default of the contractor or any of the contractor's persons, the employer or any of the employers persons, or any statutory undertaker.

11.2.2   Excepted Risk

The Excepted Risks are as defined under clause 6.1 of the Sub-contract conditions (i.e. ionising radiation or contamination by radioactivity, etc., pressure waves caused by aircraft, etc., and any act of terrorism that is not within the Terrorism Cover required to be taken out and maintained under the Main Contract).

11.2.3   Terrorism cover

Terrorism cover, being insurance provided by a Joint Names Policy under the Main Contract conditions against loss or damage to work executed and Site Materials or to an existing structure and/or its contents caused by or resulting from terrorism.

11.2.4   Insurance

In addition to the Sub-contractor's obligation to indemnify the contractor, the Sub-contractor is also to take out and maintain insurance in respect of claims arising under this liability (as clause 6.5.1 of the Sub-contract conditions).

The fact that the Sub-contractor is required to take out insurance does not affect or lessen the Sub- contractor's obligation to indemnify the contractor under clause 6.2.

The financial level of the insurance cover required under clause 6.5 is to be entered under item 13 of the Sub-contract Particulars in the Sub-contract Agreement. That said, insurance cover is to be for any one occurrence or series of occurrences arising out of one event.

11.2.5   Evidence of insurance

Under clause 6.5.3 of the Sub-contract conditions, the contractor has the right to require the Sub-contractor to produce evidence that he or she has taken out the insurances required by clause 6.5.1 and that they are being maintained. Clause 6.5.4 adds that if the Sub-contractor does not take out or maintain the insurance as required by clause 6.5.1, then the contractor may take out an insurance to cover the Sub-contractor's default, and the cost of doing so shall either be deducted from any monies due or to become due to the Sub-contractor under the Sub-contract or shall be recoverable from the Sub-contractor as a debt.

11.3   The sub-contractor's liability for injury or damage to property

11.3.1   Indemnity

The Sub-contractor is liable for and shall indemnify the contractor against any expense, liability, loss, claim or proceedings in respect of any loss, injury or damage whatsoever to any property real or personal in so far as such loss, injury or damage arises out of or in the course of or by reason of the carrying out of the Sub-contract works (as clause 6.3 of the Sub-contract conditions – refer to sections 11.2.2 and 11.2.3 of this chapter).

Clause 6.5.2 notes that this indemnity does not include for loss, injury or damage due to the effect of an Excepted Risk as defined under clause 6.1 of the Sub-contract conditions.

In line with clause 6.3 of the Sub-contract conditions, the Sub-contractor is only liable where the said loss, injury or damage is due to any negligence, breach of statutory duty, omission or default of the Sub-contractor or any of the Sub-contractor's persons.

Clause 6.4 of the Sub-contract conditions makes it clear that the above liability and indemnity shall not include any liability or indemnity in respect of loss or damage to the main contract works (including the Sub-contract works) and/or to any site materials (including the Sub-contract site materials) by any of the Specified Perils (as defined under clause 6.1 of the Sub-contract conditions).

11.3.2   Specified Perils

Fire, lightning, explosion, storm, tempest, flood, escape of water, earthquake, aircraft or other aerial devices or articles dropped therefrom, riot and civil commotion, but excluding Excepted Risks, even if this loss or damage is caused by the negligence, breach of statutory duty, omission or default of the Sub-contractor or any of the Sub-contractor's persons for the period up to and including the relevant terminal date (as defined under clause 6.1 of the Sub-contract conditions).

11.3.3   Insurance

In addition to the Sub-contractor's obligation to indemnify the contractor, the Sub-contractor is also to take out and maintain insurance in respect of claims arising under this liability (as clause 6.5.1 of the Sub-contract conditions).

Clause 6.5.1 makes it clear that such insurance is without prejudice to the Sub-contractor's obligations to indemnify the contractor under clause 6.3 of the Sub-contract conditions.

11.4   The sub-contractor's liability for loss or damage to the sub-contract works

11.4.1   Insurance

In respect of the loss or damage to the Sub-contract works and/or to any of the Sub-contract site materials by any of the Specified Perils (see section 11.3.2 of this chapter), clause 6.5.1 of the Sub-contract conditions makes it clear that the Sub-contractor is not obliged to take out and maintain insurance for this eventuality. The reason for this is because the Sub-contract works (when they form part of the main contract works) and the Sub-contract site materials (when they form part of the main contract site materials) are covered by the main contract joint names policy for loss or damage by the Specified Perils up to and including the terminal date (as defined under clause 6.1 of the sub-contract conditions).

However, the Sub-contractor needs to particularly note that the main contract joint names policy referred to above is only in respect of the Specified Perils. Other risks such as subsidence, impact, theft or vandalism (but not including Excepted Risks [see Sections 11.2.2 and 11.2.3 of this chapter]) are not covered. Also, the main contract joint names policy does not cover the Sub-contract works or the Sub-contract site materials when they are still in the custody and control of the Sub-contractor.

In this regard, it should be noted that the parties are to list under item 14 of the Sub-contract Particulars those elements of the Sub-contract works that the contractor is prepared to regard as fully, finally and properly incorporated into the main contract works prior to practical completion of the Sub-contract works or section as applicable, and also to detail the extent to which each of the listed elements needs to be carried out to achieve the status of being fully, finally and properly incorporated into the main contract works. The provision of this information reduces the probability of a dispute arising as to whether the Sub-contract works and/or the Sub-contract site materials are still in the custody and control of the Sub-contractor or not.

11.4.2   Existing structures and works therein

It should be noted that the joint names policy referred to in paragraph C.1 of insurance option C in schedule 3 of the main contract conditions does not provide any cover to Sub-contractors. This particular joint names policy relates to the insurance by the employer of existing structures and works in extensions to them. The Sub-contractor will be liable for any loss or damage caused to existing structures and/or their contents by the negligence, omission or default of the Sub-contractor or any person for whom he or she is responsible.

Such liability is a third-party liability for which the Sub-contractor is expressly liable under clause 6.3 of the Sub-contract conditions and against which he or she is required to insure under clause 6.5.1.

11.5   What Specified Perils insurance cover, in respect of loss or damage to works and site materials, does the sub-contractor obtain under the joint names all risks policies?

The cover that the Sub-contractor obtains from the all risks policy taken out under the main contract is set out under clause 6.6.1 of the Sub-contract conditions. This cover is limited to loss or damage to the Sub-contract works and the Sub-contract site materials by the Specified Perils (as defined under clause 6.1 of the sub-contract conditions).

The cover which the Sub-contractor obtains from the all risks policy is provided for by clause 6.9 of the main contract conditions. The Sub-contractor has the right (under clause 6.6.2 of the Sub-contract conditions) to require the contractor to produce evidence that he or she has taken out the insurances required by clause 6.6.1 and that they are being maintained.

In the case where the employer is a local authority and insurance option B or C applies (under the main contract), the contractor's obligation is limited to the production of a certificate certifying that terrorism cover is being provided. The reason for this exclusion is because the main contract conditions make it clear that if the employer is a local authority, then evidence of insurance does not need to be provided to the contractor.

Clause 6.6.3 of the Sub-contract conditions adds that if the contractor does not take out or maintain the insurance as required by clause 6.6.1, then the Sub-contractor may take out an insurance to cover the contractor's default, and the cost of doing so shall either be taken into account in the calculation of the final Sub-contract sum or shall be recoverable from the contractor as a debt.

11.6   What is the sub-contractor's liability for damage to the sub-contract works?

11.6.1   Damage to the sub-contract works

The Sub-contractor's responsibility in the event that there is damage to the Sub-contract works is set out under clause 6.7 of the Sub-contract conditions.

The Sub-contractor is responsible for the cost of restoration of the Sub-contract works and replacement or repair of any sub-contract site materials that are lost or damaged before the terminal date and of the removal and disposal of any resultant debris except to the extent that the loss or damage to the Sub-contract works or sub-contract site materials is due to any of the Specified Perils, or any Excepted Risk (as defined under clause 6.1 of the sub-contract conditions) or any negligence, breach of statutory duty, omission or default of the contractor, or any of the contractor's persons, or of the employer, or of any of the employer's persons or of any statutory undertaker executing work solely in pursuance of his or her statutory rights or obligations.

11.6.2   Materials fully and finally incorporated

Clause 6.7.2 of the sub-contract conditions makes it clear that if, during the progress of the sub-contract works, sub-contract materials or goods have been fully and finally incorporated into the main contract works before the terminal date, then the sub-contractor is only responsible for the cost of restoration of such work lost or damaged and the removal and disposal of any debris to the extent that such loss or damage is caused by the negligence, breach of statutory duty, omission or default of the sub-contractor or any of the sub-contractor's persons.

Clause 6.7.8 of the sub-contract conditions notes that, for the purposes of clause 6.7.2 only, materials and goods forming part of the sub-contract works shall be deemed to have been fully, finally and properly incorporated into the main contract works when in each case they have been completed by the sub-contractor to the extent indicated or referred to in the section of the sub-contract particulars that deals with the incorporation of the sub-contract works into the main contract works.

11.6.3   Notices

In the event that any loss or damage is occasioned to the sub-contract works or the sub-contract site materials, by whatever means, then, upon its occurrence or later discovery, the sub-contractor is to forthwith (i.e. immediately) give notice in writing to the contractor of its extent, nature and location.

11.6.4   Directions and variations

The contractor is then to issue his or her directions, and if those directions are such, the sub-contractor is, with due diligence, to restore the lost or damaged sub-contract works, and/or replace or repair any lost or damaged sub-contract site materials, remove and dispose of any debris, and proceed with the carrying out of the sub-contract works.

Where, in line with the above provisions, the sub-contractor is not responsible for the cost of compliance, such compliance shall be treated as a variation.

It must be noted, however, that the occurrence of loss or damage affecting the sub-contract works occasioned by any of the Specified Perils (as defined under clause 6.1 of the sub-contract conditions) shall be disregarded in computing any amounts payable to the sub-contractor under the sub-contract, although it may be payable as part of an insurance claim under the joint names all risks policy.

11.6.5 Payment from insurance

Where monies are to be paid out by the insurer under that joint names all risks policy, the sub-contractor shall not object to the payment of the monies to the employer (as clause 6.7.5 of the sub-contract conditions).

11.6.6 Terminal date

Clause 6.7.6 makes it clear that, except in the case of loss or damage caused by the negligence, breach of statutory duty, omission or default of the sub-contractor or any of the sub-contractor's persons, the sub-contractor shall not be responsible for loss or damage to the sub-contract works occurring after the terminal date, which is defined under clause 6.1 of the sub-contract conditions as being the date of practical completion of the sub-contract works or the date of termination of the sub-contractor's employment under the sub-contract, whichever occurs first.

11.6.7 Defects in sub-contract works

Finally, clause 6.7.7 of the sub-contract conditions states that nothing in clause 6.7 shall in any way modify the sub-contractor's obligations in regard to defects in the sub-contract works.

11.7 What are the employer's options where terrorism cover is not available?

11.7.1 Cessation of terrorism cover

As noted in clause 6.8 of the sub-contract conditions, if the insurers named in the joint names policy referred to in clause 6.6 notify the contractor or the employer that terrorism cover will cease on a particular date (and will no longer be available), the contractor is to immediately inform the sub-contractor.

11.7.2 Decision by employer

When the employer has notified the contractor of his or her decision (in the light of this lack of cover) either to continue with the main contract works or to terminate the contractor's employment under the main contract, the contractor is to notify the sub-contractor accordingly.

11.7.3 Termination of contractor's employment

If the employer gives notice terminating the contractor's employment under the main contract, then upon and from the date stated by the employer in his or her notice to the contractor, the sub-contractor's employment under the sub-contract shall terminate, in which case the provisions of clause 7.11 of the sub-contract conditions shall apply (but excluding for the recovery by the sub-contractor of any direct loss and/or damage caused to the sub-contractor by the termination).

11.7.4 Terrorism cover ceases

If the employer does not give notice terminating the contractor's employment under the main contract, then upon and from the date that terrorism cover will cease:

- if the sub-contract work that has been executed and/or the sub-contract site materials suffer physical loss or damage caused by terrorism, the sub-contractor shall, with due diligence, restore lost or damaged work, repair or replace any lost or damaged site materials, remove and dispose of any debris and proceed with the carrying out of the sub-contract works;
- the restoration, replacement or repair of such loss or damage, and any removal and the disposal of debris shall be treated as a variation with no reduction in any amount payable to the sub-contractor by reason of any act or neglect of the sub-contractor or of any of his or her own sub-contractors (i.e. any sub-subcontractors) which may have contributed to the physical loss or damage;
- where insurance option C applies, the requirement that the sub-contract works continue to be carried out shall not be affected by any loss or damage to the existing structures and/or their contents caused by terrorism.

11.8 Is the contractor responsible for damage caused to the sub-contractor's plant, etc.?

The contractor is only responsible for the loss of or damage to temporary works, plant tools, equipment or other property belonging to or provided by the sub-contractor, or the sub-contractor's persons, or to any materials or goods of the sub-contractor which are not sub-contract site materials when such loss or damage is due to any negligence, breach of statutory duty, omission or default of the contractor or any of the contractor's persons.

In all other instances, the sub-contractor is responsible, and the sub-contractor needs to ensure that his or her insurance arrangements with his or her own sub-contractors (i.e. the sub-subcontractors) are compatible with this liability.

11.9 Is the sub-contractor required to take out professional indemnity insurance?

11.9.1 Professional indemnity insurance cover

Clause 6.10 of the sub-contract conditions places an obligation on the sub-contractor to insure in respect of professional indemnity (PI) insurance.

The sub-contractor is required to take out this insurance of the type and with the level of cover of at least that stated in item 15 of the Sub-contract Particulars as contained in the Sub-contract agreement.

If no level of cover is stated there, then *no PI insurance is required*. This is an important point to note.

11.9.2   Commercially reasonable rates

Clause 6.10.2 of the sub-contract conditions notes that, provided that it remains available at commercially reasonable rates, the sub-contractor is to maintain the PI insurance until the expiry of the period stated under item 15 of the sub-contract particulars from the date of practical completion of the sub-contract works.

Normally, where the sub-contract is executed as a deed, the period will be 12 years, but if the sub-contract is executed under hand, the period should be 6 years.

11.9.3   Evidence of insurance

Clause 6.10.3 of the sub-contract conditions notes that the sub-contractor is required to provide documentary evidence that the PI insurance is being maintained, as and when the contractor reasonably requires. If the PI insurance is not available at commercially reasonable rates, clause 6.11 requires the sub-contractor to notify the contractor immediately so that the contractor and sub-contractor can discuss the means of best protecting their respective positions in the absence of such PI insurance.

11.10   When and how does the Joint Fire Code apply?

Clause 6.12 of the sub-contract conditions notes that the Joint Fire Code only applies where the main contract particulars state that the Joint Fire Code applies, and clause 6.13 notes that where the Joint Fire Code applies, the parties shall comply with the Joint Fire Code; the contractor is to ensure compliance by all the contractor's persons and the sub-contractor shall ensure such compliance by all the sub-contractor's persons.

11.10.1   What happens if there is a breach of the Joint Fire Code?

Clause 6.14 of the sub-contract conditions sets out the procedure to be followed in the event of a breach of the Joint Fire Code. This procedure being:

- If there is a breach of the Joint Fire Code, the insurers may require the contractor to carry out remedial measures.
- If the insurers do require the contractor to carry out remedial measures, the contractor is to send a copy of the notice requiring the remedial measures to the sub-contractor, forthwith.

- Upon receipt of the notice, the sub-contractor is to comply with any direction of the contractor that is reasonably necessary in respect of the sub-contract works to correct the breach of the Joint Fire Code.
- Except where the breach of the Joint Fire Code was as a result of an act, omission or default of the sub-contractor or of any of the sub-contractor's persons, then the compliance by the sub-contractor with any direction of the contractor, will be treated as a variation.
- If the breach of the Joint Fire Code was as a result of (or was partly as a result of) an act, omission or default of the sub-contractor or of any of the sub-contractor's persons, then the sub-contractor shall be liable for the appropriate additional costs incurred by the contractor for the remedial measures, and either an appropriate deduction shall be made in calculating the final sub-contract sum or it shall be recoverable by the contractor from the sub-contractor as a debt.
- If the sub-contractor, within 4 days of receipt of a direction, does not begin to comply with it or thereafter fails without reasonable cause regularly and diligently to comply with it, then the provisions of clause 3.6 of the sub-contract conditions will apply. Clause 3.6 makes it clear that if, within 7 days after receipt of a written notice from the contractor which requires a sub-contractor to comply with a direction, the sub-contractor still does not comply with that direction, then the contractor may employ and pay other persons to execute any work whatsoever which may be necessary to give effect to that direction.

In such a situation, the sub-contractor would be liable for all additional costs incurred by the contractor in connection with such employment and an appropriate deduction would either be taken into account in the calculation of the final sub-contract sum or be recoverable by the contractor from the sub-contractor as a debt.

11.10.2   Amendments/Revisions to the Joint Fire Code

Clause 6.15 of the sub-contract conditions makes it clear that if the Joint Fire Code is amended after the sub-contract base date, and if that amended code is to be applied in respect of the main contract works, the cost, if any, of compliance by the Sub-contractor with any such amendment shall be added into the calculations of the final Sub-contract sum.

11.11   Equivalent sub-contract provisions

In the table below, the equivalent sub-contract provisions to that within the SBCSub/D/A or SBCSub/D/C in the text above are listed for the SBCSub/A or SBCSub/C, DBSub/A or DBSub/C, ICSub/A or ICSub/C and ICSub/D/A or ICSub/D/C.

SBCSub/D/A or SBCSub/D/C	SBCSub/A or SBCSub/C	DBSub/A or DBSub/C	ICSub/A or ICSub/C	ICSub/D/A or ICSub/D/C
Sub-contract Particulars Item 13	Sub-contract Particulars Item 13	Sub-contract Particulars Item 13	Sub-contract Particulars Item 12	Sub-contract Particulars Item 12
Sub-contract Particulars Item 14	Sub-contract Particulars Item 14	Sub-contract Particulars Item 14	Sub-contract Particulars Item 13	Sub-contract Particulars Item 13
Sub-contract Particulars Item 15	Not applicable to this Sub-contract as it relates to design work to be carried out by the Sub-contractor, and Sub-contract design is not provided for under this Sub-contract	Sub-contract Particulars Item 14	Not applicable to this Sub-contract as it relates to design work to be carried out by the Sub-contractor, and Sub-contract design is not provided for under this Sub-contract	Sub-contract Particulars Item 12
Clause 3.6	Clause 3.6	Clause 3.6	Clause 3.6	Clause 3.6
Clause 6.1	Clause 6.1	Clause 6.1	Clause 6.1	Clause 6.1
Clause 6.2	Clause 6.2	Clause 6.2	Clause 6.2	Clause 6.2
Clause 6.3	Clause 6.3	Clause 6.3	Clause 6.3	Clause 6.3
Clause 6.4	Clause 6.4	Clause 6.4	Clause 6.4	Clause 6.4
Clause 6.5	Clause 6.5	Clause 6.5	Clause 6.5	Clause 6.5
Clause 6.5.1	Clause 6.5.1	Clause 6.5.1	Clause 6.5.1	Clause 6.5.1
Clause 6.5.2	Clause 6.5.2	Clause 6.5.2	Clause 6.5.2	Clause 6.5.2
Clause 6.5.3	Clause 6.5.3	Clause 6.5.3	Clause 6.5.3	Clause 6.5.3
Clause 6.5.4	Clause 6.5.4	Clause 6.5.4	Clause 6.5.4	Clause 6.5.4
Clause 6.6	Clause 6.6	Clause 6.6	Clause 6.6	Clause 6.6
Clause 6.6.1	Clause 6.6.1	Clause 6.6.1	Clause 6.6.1	Clause 6.6.1
Clause 6.6.2	Clause 6.6.2	Clause 6.6.2	Clause 6.6.2	Clause 6.6.2
Clause 6.6.3	Clause 6.6.3	Clause 6.6.3	Clause 6.6.3	Clause 6.6.3
Clause 6.7	Clause 6.7	Clause 6.7	Clause 6.7	Clause 6.7
Clause 6.7.2	Clause 6.7.2	Clause 6.7.2	Clause 6.7.2	Clause 6.7.2
Clause 6.7.5	Clause 6.7.5	Clause 6.7.5	Clause 6.7.5	Clause 6.7.5
Clause 6.7.6	Clause 6.7.6	Clause 6.7.6	Clause 6.7.6	Clause 6.7.6
Clause 6.7.7	Clause 6.7.7	Clause 6.7.7	Clause 6.7.7	Clause 6.7.7
Clause 6.7.8	Clause 6.7.8	Clause 6.7.8	Clause 6.7.8	Clause 6.7.8
Clause 6.8	Clause 6.8	Clause 6.8	Clause 6.8	Clause 6.8
Clause 6.9	Clause 6.9	Clause 6.9	Clause 6.9	Clause 6.9
Clause 6.10	Not applicable to this Sub-contract as it relates to personal indemnity insurance for design work to be carried out by the Sub-contractor, and Sub-contract design is not provided for under this Sub-contract	Clause 6.10	Not applicable to this Sub-contract as it relates to personal indemnity insurance for design work to be carried out by the Sub-contractor, and Sub-contract design is not provided for under this Sub-contract	Clause 6.14

SBCSub/D/A or SBCSub/D/C	SBCSub/A or SBCSub/C	DBSub/A or DBSub/C	ICSub/A or ICSub/C	ICSub/D/A or ICSub/D/C
Clause 6.10.2	Not applicable to this Sub-contract as it relates to personal indemnity insurance for design work to be carried out by the Sub-contractor, and Sub-contract design is not provided for under this Sub-contract	Clause 6.10.2	Not applicable to this Sub-contract as it relates to personal indemnity insurance for design work to be carried out by the Sub-contractor, and Sub-contract design is not provided for under this Sub-contract	Clause 6.14.2
Clause 6.10.3	Not applicable to this Sub-contract as it relates to personal indemnity insurance for design work to be carried out by the Sub-contractor, and Sub-contract design is not provided for under this Sub-contract	Clause 6.10.3	Not applicable to this Sub-contract as it relates to personal indemnity insurance for design work to be carried out by the Sub-contractor, and Sub-contract design is not provided for under this Sub-contract	Clause 6.14.3
Clause 6.11	Not applicable to this Sub-contract as it relates to personal indemnity insurance for design work to be carried out by the Sub-contractor, and Sub-contract design is not provided for under this Sub-contract	Clause 6.11	Not applicable to this Sub-contract as it relates to personal indemnity insurance for design work to be carried out by the Sub-contractor, and Sub-contract design is not provided for under this Sub-contract	Clause 6.15
Clause 6.12	Clause 6.12	Clause 6.12	Clause 6.10	Clause 6.10
Clause 6.13	Clause 6.13	Clause 6.13	Clause 6.11	Clause 6.11
Clause 6.14	Clause 6.14	Clause 6.14	Clause 6.12	Clause 6.12
Clause 6.15	Clause 6.15	Clause 6.15	Clause 6.13	Clause 6.13
Clause 7.11	Clause 7.11	Clause 7.11	Clause 7.11	Clause 7.11

12 Termination of Sub-contract

12.1 Introduction

Obviously, the contractor and the sub-contractor enter into the sub-contract with the intention of each performing their respective obligations, which are primarily aimed at the sub-contractor providing the sub-contract works and the contractor providing the required payment as prescribed.

In the vast majority of cases, this is achieved.

Regrettably, however, circumstances may arise when this is not possible and the contract is brought to a premature end.

Termination of a contract is potentially complex and a high-risk area. It is recommended that if a party is seeking to terminate a contract, it would be well advised to seek legal advice before taking any steps to terminate.

Against this background, in this chapter, the following matters will be dealt with, using the sub-contract clause references, etc. contained in the SBCSub/D/A or SBCSub/D/C (as appropriate). Whilst it is clearly beyond the scope of this book to review every nuance of the other sub-contract forms under consideration, the equivalent provisions (where applicable) within the SBCSub/A or SBCSub/C, the DBSub/A or DBSub/C, the ICSub/A or ICSub/C and the ICSub/D/A or ICSub/D/C are given at the table at the end of this chapter. It must be emphasised that before considering a particular issue, the actual terms of the appropriate edition of the relevant sub-contract should be reviewed by the reader (and/or legal advice should be sought as appropriate) before proceeding with any action/inaction in respect of the sub-contract in question.

In respect of the above sub-contracts, the word 'termination' is used in their respective sub-contract conditions. It is to be noted, however, that the sub-contract's termination provisions do not terminate the sub-contract itself but the sub-contractor's employment under the sub-contract. This is because, depending on the circumstances of their operation, the sub-contract still provides for various ongoing rights and obligations upon the parties after the termination of the sub-contractor's employment has occurred.

In this chapter the following matters are dealt with:

- How can a contract come to an end?
- What can cause a sub-contract to be terminated?
- What is a breach of contract?

The JCT 2011 Building Sub-contracts, First Edition. Peter Barnes and Matthew Davies.
© 2016 John Wiley & Sons, Ltd. Published 2016 by John Wiley & Sons, Ltd.

- Will any breach of contract enable one to terminate the sub-contract?
- What is a repudiatory breach at common law?
- Why have the termination provisions in the contract if the sub-contract can be terminated one common law?
- What is the effect of a sub-contract being terminated at common law?
- Are the effects of a sub-contract being terminated under common law or by contractual provisions the same?
- Where can the contractual termination provisions be located in the sub-contract?
- Are the party's common law termination rights preserved under the sub-contract?
- What reasons give the contractor a right to terminate the sub-contractor's employment under the JCT sub-contract conditions?
- What is deemed to be default by the sub-contractor under clause 7.4.1 of SBCSub/D/C?
- What happens when a sub-contractor commits a specified default noted at clause 7.4.1 of SBCSub/D/C?
- Must all notices by the contractor or the sub-contractor referred to under Section 7 'Termination' of the sub-contract conditions be given in accordance with clause 1.7.4?
- What does insolvency of the sub-contractor mean?
- What happens when the sub-contractor becomes insolvent?
- What does corruption entail?
- What are the consequences of the contractor terminating the sub-contractor's employment?
- What reasons give the sub-contractor the right to terminate his or her employment under the JCT sub-contract conditions?
- What is deemed to be default by the contractor under clause 7.8.1 of SBCSub/D/C?
- What happens when a contractor commits a specified default as recorded in clause 7.8.1 of SBCSub/D/C?
- What does insolvency of the contractor mean?
- What happens when the contractor becomes insolvent?
- What are the consequences of the sub-contractor terminating his or her own employment?
- If a sub-contractor's employment is terminated for any reason, can it subsequently be reinstated?

12.2   How can a contract come to an end?

There are essentially four ways in which a contract can come to an end:

- Performance
- Agreement
- Frustration
- Termination for breach

For the purposes of this book, it is the last category (termination) that is being considered.

In respect of the sub-contracts that are being considered in this book, termination can arise in two ways. These are:

- Termination pursuant to a party's common law rights
- Termination using a contractual provision contained within the sub-contract

12.3   What can cause a sub-contract to be terminated?

If there is a breach of an important condition of the sub-contract, then a party has a right to terminate the sub-contract in certain permitted circumstances. These are:

- Termination at common law: For this to apply, a 'fundamental' breach of contract going to the very root of the contract must have been committed. This is called a repudiatory breach by lawyers. Caution is required here as not all breaches of contract will qualify as repudiatory. This generates uncertainty and risk in operating the common law right to terminate a contract. A party terminating a sub-contract where the common law right does not actually exist is most likely committing, by this act, a repudiatory breach themselves.
- Termination under the contract provisions: In an endeavour to avoid the perceived uncertainty and risk of the common law position, the sub-contracts provide the right to terminate the sub-contractor's employment under the sub-contract in certain prescribed situations. To be effective, it is imperative that a specified termination ground exists and that the procedure is followed precisely. Where this does not occur, the termination will be wrongful.

12.4   What is a breach of contract?

A breach of contract is committed when a party, without lawful excuse, fails or refuses to perform what is due from him or her under the contract or performs defectively or incapacitates himself or herself from performing what is due.

12.5   Will any breach of contract enable one to terminate the sub-contract?

No.

When a term of a contract is breached, various remedies may be are available, but generally, the usual one is compensation to the innocent party via the receipt of damages.

It is important to recognise that most breaches do not prevent the ongoing performance of the contract and can be compensated by damages alone.

For example, the sub-contractor's interim payment no. 9 is paid 3 weeks late. This is a breach and can be compensated by damages (interest). On the facts it is not a fundamental (repudiatory) breach going to the root of the contract and would not ordinarily be considered grounds at common law enabling a party to terminate the contract (although the contractual termination may provide that this is a specified default ground – one would need to check).

For a breach of contract to qualify for the right of the innocent party to terminate the contract, it must be either:

■ a repudiatory (fundamental) breach in order for the common law rights to terminate to apply; or

■ a stated specified default ground in order for the contractual termination provisions to apply.

12.6   What is a repudiatory breach at common law?

■ A repudiatory breach is a serious, fundamental breach of contract that goes to the root of the contract. A repudiatory breach is a breach whereby one party clearly indicates (by words or by conduct) that he or she no longer intends to be bound by the terms of the sub-contract, or perhaps where the default of a party has rendered himself or herself unable to perform his or her outstanding contractual obligations (e.g. where a contractor needs to be NHBC registered but he or she is removed from the NHBC register); or

■ an anticipatory breach where one party states that he or she will not be carrying out his or her obligations under the sub-contract before the time for carrying out those obligations has actually arrived.

Once either a repudiatory breach or an anticipatory breach has occurred, then the innocent party may either affirm the sub-contract or elect to accept the repudiatory breach or the anticipatory breach as appropriate and then terminate the sub-contract.

It is important to note that the repudiatory breach or the anticipatory breach needs to be accepted, because, as noted in the *White & Carter* v. *McGregor*[1] case: 'Repudiation by one party standing alone does not terminate the contract. It takes two to end it, by repudiation on the one side, and acceptance of the repudiation on the other'.

12.7   Why have termination provisions in the contract if the sub-contract can be terminated at common law?

It is not unusual for there to be arguments under common law about whether the condition that was breached was important enough to allow a party to terminate the contract in the first place. These can escalate quite quickly into expensive dispute and legal proceedings.

---

[1] *White & Carter (Councils) Ltd* v. *McGregor* [1962] AC 413, HL.

In respect of contracts that contain a contractual termination clause, this problem is overcome because those breaches that will allow contractual termination to take place are defined in the contract itself. Therefore, in a contractual termination, it will not be necessary to show that the breach was of a sufficiently fundamental nature to warrant termination; all that will need to be shown is that the breach relied upon has been defined within the contract as being a breach that allows contractual termination to take place.

However, for a contractual termination to be successful, it will be necessary for the contract procedures for notices, etc. to be followed meticulously, otherwise the party pursuing the contractual termination route may find himself or herself accused of a repudiatory breach.

12.8 What is the effect of a sub-contract being terminated at common law?

It is reasonably well established that when a sub-contract is terminated by a party because of a breach of an important condition by the other party, the innocent party is released from all further performance of his or her obligations but he or she is entitled to recover full damages arising from the breach. Therefore, termination of the sub-contract in such a situation actually means that a party's employment under the sub-contract is terminated, rather than that the sub-contract is terminated per se. Although strictly speaking, it is unnecessary to state this fact, it is normal practice to say that a party's 'employment under the sub-contract' has been terminated, and it is suggested that this practice should be followed for the avoidance of any possible doubt.

12.9 Are the effects of a sub-contract being terminated under common law or by contractual provisions the same?

Generally, the rights and remedies under a common law termination may well be different from those under a contractual termination.

For example, under a common law termination (unlike most contractual termination clauses), the innocent party may decide not to complete the project but may still claim damages.

On the other hand, under contractual termination, the innocent party is frequently entitled to seize materials to complete the works, which would not be the case under a common law termination.

Also, under a contractual termination, the consequences arising from the termination are clearly set out, and this may include the payment to the innocent party of all sums due within a relatively short timescale.

In the case of common law termination, a high degree of freedom and flexibility of remedy is available, but the innocent party has the burden of establishing that the breach of sub-contract was sufficiently fundamental to justify the termination in law.

A fundamental point to note is that, unless specifically excluded, it is generally the case that the right to common law termination coexists with any contractual termination provisions. In other words, unless specifically excluded, a party's common law termination rights are preserved irrespective of any contractual termination provisions.

12.10 Where can the contractual termination provisions be located in the Sub-contract?

Section 7 of SBCSub/D/C deals with termination.

12.11 Are the party's common law termination rights preserved under the Sub-contract?

Yes. Clause 7.3.1 of SBCSub/D/C makes it clear that 'the provisions of clause 7.4 to 7.8, 7.10 and 7.11 are without prejudice to any other rights and remedies which the contractor or the Sub-contractor may possess.' This clause therefore effectively confirms that the parties' common law termination rights are preserved.

12.12 What reasons give the contractor a right to terminate the sub-contractor's employment under the JCT Sub-contract Conditions?

The contractor may terminate the sub-contractor's employment under the sub-contract conditions for three principal reasons:

- Default by the sub-contractor (clause 7.4.1 refers)
- Insolvency of the sub-contractor (clause 7.5 refers)
- Corruption (clause 7.6 refers)

The sub-contractor's employment under the sub-contract will also occur on termination of the contractor's employment under the Main Contract (clause 7.9 refers)

12.13 What is deemed to be default by the sub-contractor under clause 7.4.1 of SBCSub/D/C?

The specified defaults (which, to be defaults, must be committed before practical completion of the sub-contract works) are:

- The sub-contractor without reasonable cause wholly or substantially suspends the carrying out of the sub-contract works or the design of the sub-contractor's designed portion (as clause 7.4.1.1 of SBCSub/D/C).

This ground for termination does not apply where the sub-contractor has reasonable cause for suspending the carrying out of the sub-contract works.

Therefore, a valid suspension because of non-payment correctly undertaken in accordance with clause 4.11 of the sub-contract conditions would be a reasonable cause for suspending the carrying out of the sub-contract works, as this would not be a specified default.

■ The sub-contractor fails to proceed regularly and diligently with the sub-contract works or the design of the sub-contractor's designed portion (as clause 7.4.1.2 of SBCSub/D/C).

The word 'regularly' suggests a requirement to attend for work on a regular daily basis with sufficient men, materials and plant to have the physical capacity to progress the works substantially in accordance with the sub-contract obligations.

'Diligently' adds the concept of the need to apply the above physical capacity industriously and efficiently towards that same end.

Taken together, the obligation upon the sub-contractor is essentially to proceed continuously, industriously and efficiently with appropriate physical resources so as to progress the works steadily towards completion, substantially in accordance with the contractual requirements as to time, sequence and quality of work.

■ The sub-contractor refuses or neglects to comply with a written direction from the contractor requiring him or her to remove any work, materials or goods not in accordance with the sub-contract, and by such refusal or neglect, the main contract works are materially affected (as clause 7.4.1.3 of SBCSub/D/C).

■ The sub-contractor fails to comply with clause 3.1 (i.e. non-assignment) or 3.2 (consent to be obtained before sub-letting) of the sub-contract conditions (as clause 7.4.1.4 of SBCSub/D/C).

■ The sub-contractor fails to comply with the CDM Regulations (as clause 7.4.1.5 of SBCSub/D/C).

12.14 What happens when a sub-contractor commits a specified default noted at clause 7.4.1 of SBCSub/D/C?

The basic process under clause 7.4 of SBCSub/D/C that is followed is:

■ The sub-contractor defaults in a way stipulated in the sub-contract (as noted in clauses 7.4.1.1–clause 7.4.1.5).

■ It is important to note that for clause 7.4 to apply, the default must be in the manner listed as being a 'specified default' in clause 7.4.1.

■ The contractor gives a notice to the sub-contractor specifying the default (clause 7.4.1 refers).

■ If the sub-contractor continues a specified default for 10 calendar days (excluding any public holiday days) after receipt of the contractor's notice, then the contractor may on or within 21 calendar days (excluding any public holiday days) from the expiry of the initial 10–calendar day notice period, issue a further notice terminating the sub-contractor's employment under the sub-contract (clause 7.4.2 refers).

■ If the contractor does not give the further notice referred to in clause 7.4.2 (for any reason), but the sub-contractor repeats a specified default that has been the subject of an earlier notice under clause 7.4.1 (whether previously repeated or not), then, upon or within a reasonable time after such repetition, the contractor may by notice to the sub-contractor terminate the sub-contractor's employment under the sub-contract (clause 7.4.3 refers).

■ It is important to note that clause 7.2.3 provides that the notices given above must be given in accordance with clause 1.7.4. of SBCSub/D/C.

12.15   Must all notices either by the Contractor or the Sub-contractor referred to under Section 7 'Termination' of the sub-contract conditions be given in accordance with clause 1.7.4?

Yes. Clause 7.2.3 of SBCSub/D/C clearly states that all notices that are given under section 7 *must* be given in accordance with clause 1.7.4. of SBCSub/D/C.

■ Clause 1.7.4 provides that such notices must be delivered by hand or sent by Recorded Signed or Special Delivery post. Where the notice is given by Recorded Signed or Special Delivery post, it shall, subject to proof to the contrary, be deemed to have been received on the second business day after the date of posting.

■ The words 'subject to proof to the contrary' are important, because this means that (subject to proof) a notice could have been actually received prior to the second business day after the date of posting.[2]

■ Termination shall take effect on receipt of the relevant notice (as clause 7.2.2 of SBCSub/D/C).

■ Notice of termination of the sub-contractor's employment shall not be given unreasonably or vexatiously (as clause 7.2.1 of SBCSub/D/C).

12.16   What does insolvency of the sub-contractor mean?

In line with clause 7.1 of the sub-contract conditions (SBCSub/D/C), and *for the purposes of the sub-contract conditions only*, a sub-contractor is insolvent if

(1) a Party which is a company becomes insolvent:
   (a) when it enters administration within the meaning of schedule B1 to the Insolvency Act 1986;
   (b) on the appointment of an administrative receiver or a receiver or manager of its property under chapter 1 of part III of that Act, or the appointment of a receiver under Chapter II of that part;

---

[2]*Lafarge (aggregotes) Ltd* v. *Newham Borough Council* [2005] (unreported) QBD, 24 June 2005.

      (c) on the passing of a resolution for voluntary winding up without a declaration of solvency under section 89 of that Act; or

      (d) on the making of a winding-up order under part IV or V of that Act.

(2)   a Party which is a partnership becomes insolvent:

      (a) on the making of a winding-up order against it under any provision of the Insolvency Act 1986 as applied by an order under section 420 of that Act; or

      (b) when sequestration is awarded on the estate of the partnership under section 12 of the Bankruptcy (Scotland) Act 1985 or the partnership grants a trust deed for its creditors.

(3)   a Party who is an individual becomes insolvent:

      (a) on the making of a bankruptcy order against him or her under part IX of the Insolvency Act 1986; or

      (b) on the sequestration of his or her estate under the Bankruptcy (Scotland) Act 1985 or when he or she grants a trust deed for his or her creditors.

(4)   a Party also becomes insolvent if:

      (a) he or she enters into an arrangement, compromise or composition in satisfaction of his or her debts (excluding a scheme of arrangement as a solvent company for the purposes of amalgamation or reconstruction); or

      (b) (in the case of a Party which is a partnership) each partner is the subject of an individual arrangement or any other event or proceedings referred to in clause 7.1.

Each of clauses 7.1.1–7.1.4 also includes any analogous arrangement, event or proceedings in any other jurisdiction.

Clause 7.5.2 of the sub-contract conditions requires the sub-contractor to immediately inform the contractor in writing if he or she makes any proposal, gives notice of any meeting or becomes the subject of any proceedings or appointment to any of the matters referred to above.

12.17   What happens when the sub-contractor becomes insolvent?

Clause 7.5.1 of SBCSub/D/C states that if the sub-contractor is insolvent, the contractor may at any time by notice to the sub-contractor terminate the sub-contractor's employment under the sub-contract.

It is important to note that clause 7.2.3 provides that all notices that are given under Section 7 'Termination' of SBCSub/D/C *must* be given in accordance with clause 1.7.4.

It is interesting to note that upon a sub-contractor becoming insolvent, his or her employment under the sub-contract is not automatically terminated and a notice from the contractor (which may be made at any time) is required to bring the termination of employment into effect.

However, clause 7.5.3.1 provides that the sub-contractor's obligations to carry out and complete the sub-contract works (including the design works, where applicable) are suspended from the date the sub-contractor is insolvent (regardless of the contractor having given notice).

The assumed reason why there is not automatic termination, but there is a period of suspension, is to allow a period following the happening of a specified insolvency event during which the sub-contract arrangements will remain in force while the parties attempt to agree a way forward.

In addition, clause 7.5.3.2 of the sub-contract conditions states that, as from when the sub-contractor becomes insolvent (which may be at a date earlier than when the sub-contractor's employment under the sub-contract has been terminated), the contractor may take reasonable measures to ensure that the sub-contract works and the sub-contract site materials are:

- adequately protected;
- retained on site.

The sub-contractor shall allow and shall not hinder or delay the taking of those measures.

12.18   What does corruption entail?

Clause 7.6 of SBCSub/D/C deals with corruption.

Where corruption occurs as specified therein, clause 7.6 of SBCSub/D/C permits the contractor to be entitled, by notice to the sub-contractor, 'to terminate the Sub-contractor's employment under this sub-contract or any other contract with the Contractor'.

Corruption is deemed only to be when the sub-contractor or any person employed by him or her or acting on his or her behalf has committed an offence under the Bribery Act 2010 or, where the employer is a local authority, has given any fee or reward, the receipt of which is an offence under sub-section (2) of section 117 of the Local Government Act 1972.

It is important to note that clause 7.2.3 provides that all notices that are given under Section 7 'Termination' of SBCSub/D/C *must* be given in accordance with clause 1.7.4.

12.19   What are the consequences of the contractor terminating the sub-contractor's employment?

Clause 7.7 of SBCSub/D/C deals with the consequences of termination under clauses 7.4–7.6 of SBCSub/D/C.

If the sub-contractor's employment is terminated by the contractor for any of the reasons outlined under clauses 7.4–7.6 of SBCSub/D/C, the following consequences arise:

- The contractor may employ and pay other persons to carry out and complete the sub-contract works (including the design for the sub-contractor's designed portion, where applicable) and to make good any defects of the kind referred to in clause 2.22 of SBCSub/D/C (refer to clause 7.7.1 of SBCSub/D/C).

■ The contractor (and/or the persons employed by the contractor) may enter upon and take possession of the sub-contract works and (subject to obtaining any necessary third-party consents) may use all the sub-contractor's temporary buildings, plant, tools, equipment and sub-contract site materials for those purposes (refer to clause 7.7.1 of SBCSub/D/C).

It should be noted that there may be restrictions placed on the contractor's right to use site materials, etc. of a sub-contractor who has become insolvent, because this may prejudice other creditors of the sub-contractor and may not be compliant with the mandatory equal discharge of liabilities by a liquidator or receiver, something that cannot be excluded by contract.[3]

■ When required by the contractor in writing (but not beforehand), the sub-contractor must remove or procure the removal from the site of all temporary buildings, plant, tools, equipment, goods and materials belonging to the sub-contractor or to the sub-contractor's persons (as clause 7.7.2.1 of SBCSub/D/C).

■ Where applicable, and without charge, the sub-contractor is to provide the contractor with three copies of all the sub-contractor's design documents then prepared, whether or not previously provided. This consequence is required irrespective of any request by the contractor (as clause 7.7.2.2 of SBCSub/D/C).

If required by the contractor, the sub-contractor must within 14 days of the date of termination assign to the contractor, without charge, the benefit of any agreement for the supply of materials or goods and/or for the execution of any work for the purposes of the sub-contract (as clause 7.7.2.3 of SBCSub/D/C).

Of course, the above will only be possible to the extent that:

○ the sub-contractor can legally carry out or is legally required to carry out such assignments (noting that assignment may not be legally possible in the case of the sub-contractor's insolvency);

○ the sub-contractor's own sub-contractor or supplier is prepared for the agreement to be assigned.

■ Clause 7.7.3 is important as this provides that:

○ no further sum (i.e. future payments) shall become due to the sub-contractor except for as provided for under clause 7.7.4;

○ in respect of any sums that have already become due to the sub-contractor prior to the termination of the sub-contractor's employment, then the contractor need not pay the same, provided that:

■ the contractor has given or gives a Pay Less notice as required by clause 4.10.5; or

■ if the sub-contractor becomes insolvent within the meaning of clauses 7.1.1–7.1.3 after the last date when a Pay Less notice

---

[3] *British Eagle International Air Lines v. Compagnie Nationale Air France* [1975] 1 WLR 758.

could have been given by the contractor to the sub-contractor in respect of the sum due.
- Clause 7.7.4 of SBCSub/D/C sets out the sub-contractor's rights in respect of future payment as follows:
  - Upon the completion of the sub-contract works and the making good of defects (i.e. by the contractor or by a person appointed by the contractor) or upon the termination of the contractor's employment under the main contract, whichever occurs first, the sub-contractor may apply to the contractor for payment.

The contractor needs to take no action until an application for payment is received from the sub-contractor that is compliant with the above time restriction.
  - When such an application for payment is received by the contractor, the contractor is to pay the sub-contractor the value of any work executed or goods and materials supplied by the sub-contractor to the extent not included in previous payments. Without prejudice to any other rights that the contractor may have, the contractor may deduct from the sum determined:
    - the amount of any direct loss and/or damage caused to the contractor as a result of the termination (e.g. the cost of completing the works, and other delay costs);
    - any other amount payable to the contractor under the sub-contract.

In that the amount due to the contractor exceeds the amount due to the sub-contractor, the balance is recoverable from the sub-contractor as a debt.

12.20   What reasons give the sub-contractor the right to terminate his or her employment under the JCT sub-contract conditions?

The sub-contractor may terminate his or her employment under the sub-contract conditions for three principal reasons:

- Default by the contractor (clause 7.8 of SBCSub/D/C refers)
- Termination of the main contractor's employment under the main contract (clause 7.9 of SBCSub/D/C refers)
- Insolvency of the contractor (clause 7.10 of SBCSub/D/C refers)

12.21   What is deemed to be default by the contractor under clause 7.8.1 of SBCSub/D/C?

The specified defaults are dealt with under clause 7.8.1 of SBCSub/D/C. These specified defaults are:

- The contractor without reasonable cause wholly or substantially suspends the carrying out of the main contract works (as clause 7.8.1.1 of SBCSub/D/C).

This ground for termination does not apply where the contractor has reasonable cause for suspending the carrying out of the main contract works.

■ The contractor without reasonable cause fails to proceed with the main contract works with the result that the reasonable progress of the sub-contract works is seriously affected (as clause 7.8.1.2 of SBCSub/D/C).

■ The contractor fails to make payment in accordance with the sub-contract (as clause 7.8.1.3 of SBCSub/D/C).

This is an interesting and quite onerous clause, particularly given that the sub-contractor has the express right to interest on late payments (as clauses 4.10.6 and 4.12.7 of the sub-contract conditions), and the right to suspend performance (as clause 4.11 of the sub-contract conditions) in the appropriate cases.

■ The contractor fails to comply with the CDM Regulations (as clause 7.8.1.4 of SBCSub/D/C).

It is submitted that minor or trifling failures would not be considered to be a default, given that a notice of default and/or termination of the sub-contractor's employment is not to be given unreasonably or vexatiously.

12.22   What happens when a contractor commits a specified default noted under clause 7.8.1 of SBCSub/D/C?

Clause 7.8 of SBCSub/D/C refers.

The basic process that is followed is:

■ The contractor defaults in the manner specified in clause 7.8.1 (note for clause 7.8 to apply, the default must be in the manner listed as being a 'specified default' at clause 7.8.1).

■ The sub-contractor gives a notice to the contractor specifying the default (clause 7.8.1 refers).

■ If the contractor continues a specified default for 10 calendar days (excluding any public holidays) after receipt of the sub-contractor's notice, then the sub-contractor may on or within 21 calendar days (excluding any public holidays) from the expiry of the initial 10-day notice period issue a further notice terminating the sub-contractor's employment under the sub-contract (clause 7.8.2 refers).

■ If the sub-contractor does not give the further notice referred to in clause 7.8.2, but the contractor repeats a specified default that has been the subject of an earlier notice under clause 7.8.1 (whether previously repeated or not), then, upon or within a reasonable time after such repetition, the sub-contractor may by notice to the contractor terminate the sub-contractor's employment under the sub-contract (clause 7.8.3 refers).

■ It is important to note that clause 7.2.3 provides that the notices given above must be given in accordance with clause 1.7.4.

Clause 1.7.4 provides that such notices must be delivered by hand or sent by Recorded Signed or Special Delivery post. Where the notice is given by Recorded Signed or Special Delivery post, it shall, subject to proof to the contrary, be deemed to have been received on the second business day after the date of posting. The words 'subject to proof to the contrary' are important, because this means that (subject to proof) a notice could have been actually received prior to the second business day after the date of posting.[4]

- Termination shall take effect on receipt of the relevant notice (as clause 7.2.2 of the sub-contract conditions).
- Notice of termination of the sub-contractor's employment shall not be given unreasonably or vexatiously (as clause 7.2.1 of the sub-contract conditions).

12.23   What does insolvency of the contractor mean?

Clause 7.10 of SBCSub/D/C deals with the position where the contractor is insolvent.

Clause 7.10.1 of SBCSub/D/C states that if the contractor is insolvent or makes any proposal, gives notice of any meeting or becomes the subject of any proceedings relating to any of the matters referred to in clause 7.1 of SBCSub/D/C, the contractor shall immediately notify the sub-contractor.

Clause 7.1 of the sub-contract conditions lists and defines various circumstances where a contractor (or a sub-contractor) is classified as insolvent under the sub-contract. The current categories at the time of writing are already listed under the earlier question in this chapter 'What does the insolvency of the sub-contractor mean?' (Section 12.16) and are not therefore repeated here.

Further, in any insolvency situation (or any other situation where the terms of the sub-contract need to be checked by the parties), it is imperative that the terms and definitions in the actual sub-contract entered into between the parties are referred to, as these can be updated, changed and/or amended and may not be those current at the time of writing this book.

12.24   What happens when the contractor becomes insolvent?

Clause 7.10.2 of the sub-contract conditions states that if the contractor is insolvent (as defined), then:

- the sub-contractor is entitled by notice to the contractor to terminate the sub-contractor's employment under the sub-contract.
- if, before the contractor became insolvent, the sub-contractor had already issued a notice of 'default', then the sub-contractor may issue the 'termination' notice immediately that the contractor becomes insolvent.
- if, however, before the contractor became insolvent, the sub-contractor had *not* issued a notice of 'default', then when the contractor becomes insolvent, the sub-contractor may either:
  - ○ issue a notice of termination of the sub-contractor's employment (in line with clause 7.10.2.1 of SBCSub/D/C); or
  - ○ issue a notice of default pursuant to clause 7.8.1 of SBCSub/D/C.

---

[4]*Lafarge (aggregates) Ltd* v. *Newham Borough Council* [2005] (unreported) QBD, 24 June 2005.

Whichever of these notices the sub-contractor decides to issue, clause 7.10.2.1 of the sub-contract conditions makes it clear that the sub-contractor is not to exercise his or her rights to actually terminate his or her employment under the sub-contract prior to the expiry of a period of 3 weeks (or such further periods as the parties may agree) from the date upon which the contractor became insolvent.

The reason for this 3-week period is to allow a period following the happening of a specified insolvency event during which the sub-contract arrangements will remain in force while the parties attempt to agree a way forward.

In line with clause 7.10.2.2 of SBCSub/D/C, the sub-contractor's obligations under article 2 and the sub-contract conditions to carry out and complete the sub-contract works and the design of the sub-contractor's designed portion are immediately suspended, pending termination of the sub-contractor's employment; such suspension shall, for the purposes of a relevant sub-contract event for extension of time and for the purposes of a relevant sub-contract matter, be deemed to be a default of the contractor.

12.25   What are the consequences of the sub-contractor terminating his or her own employment?

If the sub-contractor's employment is terminated by the sub-contractor for any of the following reasons:

■ Default by the contractor (clause 7.8 of SBCSub/D/C refers)
■ Termination of the main contractor's employment under the main contract (clause 7.9 of SBCSub/D/C refers)
■ Insolvency of the contractor (clause 7.10 of SBCSub/D/C refers)

then the following consequences apply as provided for by clause 7.11 of SBCSub/D/C:

■ The contractor is not required to make any further payment or release any outstanding retention other than as provided for under clause 7.11 of SBCSub/D/C.
■ In line with clause 7.11.2 of SBCSub/D/C:
   ○ clause 7.11.2.1 provides that the sub-contractor shall with all reasonable dispatch remove or procure the removal from the site of any temporary buildings, plant, tools and equipment belonging to the sub-contractor and the sub-contractor's persons, together with all goods and materials, including site materials (except only for those site materials that will be valued in the sub-contractor's concluding account and which will be paid for by the contractor (refer to clause 7.11.5 of SBCSub/D/C);
   ○ clause 7.11.2.2 also provides that without charge, the sub-contractor is to provide the contractor with three copies of all the sub-contractor's design documents then prepared as referred to in clause 2.24. This consequence is required irrespective of any request by the contractor.

- In line with clause 7.11.3 of the sub-contract conditions, the sub-contractor shall with reasonable dispatch prepare and submit to the contractor an account setting out:
  - ○ the total value of the work properly executed at the date of termination, ascertained and valued in line with the sub-contract conditions;
  - ○ The total value of any other amounts due to the sub-contractor at the date of termination, ascertained and valued in line with the sub-contract conditions, for example, any applicable fluctuation payments.

    Also, given that the next point on this list only relates to any ascertained direct loss and/or expense, these 'other amounts' would presumably also include for other loss and/or expense items claimed by the sub-contractor but not ascertained by the contractor.
  - ○ Any sums ascertained in respect of direct loss and/or expense under clause 4.19 of the sub-contract conditions.
  - ○ The reasonable cost of removal from the site of any temporary buildings, plant, tools and equipment belonging to the sub-contractor and the sub-contractor's persons, together with all goods and materials, including site materials, as appropriate.
  - ○ The cost of materials and goods (including sub-contract site materials) properly ordered for the sub-contract works for which the sub-contractor has already made payment, or for which the sub-contractor is legally liable to make payment.
  - ○ Any direct loss and/or damage caused to the sub-contractor by the termination of his or her employment. This could, of course, include, if applicable, the loss of future overheads' contribution and the loss of future profit.

In respect of the sub-contractor's recovery of direct loss and/or damage caused to the sub-contractor by the termination of his or her employment, clause 7.11.4 provides that this will *not* apply where:
  - ○ Clause 7.11.4.1: the termination of the sub-contractor's employment is due to the termination of the contractor's employment (by either the employer or the contractor) by reason of those events covered by clause 8.11.1 of the main contract conditions, unless and except in circumstances where:
    - it relates to loss or damage to the main contract works occasioned by one of the specified perils, where such loss or damage was caused by the negligence or default of the employer or any of the employer's persons.
  - ○ the termination of the sub-contractor's employment follows the employer's option in terminating the contractor's employment in the situation where there was a non-availability of insurance cover for terrorism (refer to clause 6.11.2.2 of the main contract conditions) (as clause 7.11.4.2 of the sub-contract conditions).
  - ○ the termination of the sub-contractor's employment is due to the termination of the contractor's employment (by either the employer or the contractor) because it is considered just and equitable to do so following

loss or damage to the works occasioned by any of the risks covered by paragraph C.4.4 of schedule 3 to the main contract conditions) (as clause 7.11.4.2 of the sub-contract conditions).

- The contractor shall pay to the sub-contractor the amount properly due in respect of the account submitted by the sub-contractor, without any deduction for retention, but after taking into account amounts already paid, within *28 days* of the submission of the account by the sub-contractor to the contractor.

  Payment by the contractor for any materials and goods within the account is made on the basis that the said materials and goods become the property of the contractor upon payment being made (refer to clause 7.11.5 of the sub-contract conditions).

12.26 If a sub-contractor's employment is terminated for any reason, can it subsequently be reinstated?

Clause 7.3.2 of SBCSub/D/C makes it clear that, irrespective of the grounds of termination, the sub-contractor's employment may at any time be reinstated on such terms as the parties may agree. It should be noted, however, that section 178 of the Insolvency Act 1986 gives a liquidator or a receiver the statutory right to disclaim unprofitable contracts, and this may close this avenue in certain circumstances.

12.27 Equivalent sub-contract provisions

In the table below, the equivalent sub-contract provisions to that within the SBCSub/D/A or SBCSub/D/C in the text above are listed for the SBCSub/A or SBCSub/C, DBSub/A or DBSub/C, ICSub/A or ICSub/C and ICSub/D/A or ICSub/D/C.

SBCSub/D/A or SBCSub/D/C	SBCSub/A or SBCSub/C	DBSub/A or DBSub/C	ICSub/A or ICSub/C	ICSub/D/A or ICSub/D/C
Article 2	Article 2	Article 2	Article 2	Article 2
Section 7	Section 7	Section 7	Section 7	Section 7
Clause 1.7.4	Clause 1.7.4	Clause 1.7.4	Clause 1.7.4	Clause 1.7.4
Clause 7.1	Clause 7.1	Clause 7.1	Clause 7.1	Clause 7.1
Clauses 7.2.1–7.2.3	Clauses 7.2.1–7.2.3	Clauses 7.2.1–7.2.3	Clauses 7.2.1–7.2.3	Clauses 7.2.1–7.2.3
Clause 7.3.1	Clause 7.3.1	Clause 7.3.1	Clause 7.3.1	Clause 7.3.1
Clause 7.3.2	Clause 7.3.2	Clause 7.3.2	Clause 7.3.2	Clause 7.3.2
Clause 7.4.1.1	Clause 7.4.1.1 Drafting refers to Sub-contract Works only (design is not part of the sub-contract works)	Clause 7.4.1.1 Drafting refers to Sub-contract Works only (but sub-contractor's design is part of the sub-contract works)	Clause 7.4.1.1 Drafting refers to Sub-contract Works only (design is not part of the sub-contract works)	Clause 7.4.1.1

SBCSub/D/A or SBCSub/D/C	SBCSub/A or SBCSub/C	DBSub/A or DBSub/C	ICSub/A or ICSub/C	ICSub/D/A or ICSub/D/C
Clause 7.4.1.2	Clause 7.4.1.2 Drafting refers to Sub-contract Works only (design is not part of the sub-contract works)	Clause 7.4.1.2 Drafting refers to performance of obligations (includes therefore Sub-contract Works and design)	Clause 7.4.1.2 Drafting refers to Sub-contract Works only (design is not part of the sub-contract works)	Clause 7.4.1.2
Clause 7.4.1.3	Clause 7.4.1.3	Clause 7.4.1.3	Clause 7.4.1.3	Clause 7.4.1.3
Clause 7.4.1.4	Clause 7.4.1.4	Clause 7.4.1.4	Clause 7.4.1.4	Clause 7.4.1.4
Clause 7.4.1.5	Clause 7.4.1.5	Clause 7.4.1.5	Clause 7.4.1.5	Clause 7.4.1.5
Clause 7.4.2	Clause 7.4.2	Clause 7.4.2	Clause 7.4.2	Clause 7.4.2
Clause 7.4.3	Clause 7.4.3	Clause 7.4.3	Clause 7.4.3	Clause 7.4.3
Clause 7.5.1	Clause 7.5.1	Clause 7.5.1	Clause 7.5.1	Clause 7.5.1
Clause 7.5.2	Clause 7.5.2	Clause 7.5.2	Clause 7.5.2	Clause 7.5.2
Clause 7.5.3.1	Clause 7.5.3.1 Drafting refers to Sub-contract Works only suspended (i.e. design is not part of the sub-contract works)	Clause 7.5.3.1 Drafting refers to Sub-contract Works only suspended (but design is part of the sub-contract works)	Clause 7.5.3.1 Drafting refers to Sub-contract Works only suspended (i.e. design is not part of the sub-contract works)	Clause 7.5.3.1
Clause 7.5.3.2	Clause 7.5.3.2	Clause 7.5.3.2	Clause 7.5.3.2	Clause 7.5.3.2
Clause 7.6	Clause 7.6	Clause 7.6	Clause 7.6	Clause 7.6
Clause 7.7.1	Clause 7.7.1 No reference to employing others to complete design as design is not undertaken by the Sub-contractor under this sub-contract	Clause 7.7.1 Reference to employing others to complete the sub-contract works would include, as appropriate, any incomplete sub-contractor designs	Clause 7.7.1 No reference to employing others to complete design as design is not undertaken by the Sub-contractor under this sub-contract	Clause 7.7.1
Clause 7.7.2.1	Clause 7.7.2.1	Clause 7.7.2.1	Clause 7.7.2.1	Clause 7.7.2.1
Clause 7.7.2.2	N/A	Clause 7.7.2.2	N/A	Clause 7.7.2.2
Clause 7.7.2.3	Clause 7.7.2.2	Clause 7.7.2.3	Clause 7.7.2.2	Clause 7.7.2.3
Clause 7.7.3.1	Clause 7.7.3.1	Clause 7.7.3.1	Clause 7.7.3.1	Clause 7.7.3.1
Clause 7.7.3.2	Clause 7.7.3.2	Clause 7.7.3.2	Clause 7.7.3.2	Clause 7.7.3.2
Clause 7.7.4	Clause 7.7.4	Clause 7.7.4	Clause 7.7.4	Clause 7.7.4
Clause 7.8.1.1	Clause 7.8.1.1	Clause 7.8.1.1	Clause 7.8.1.1	Clause 7.8.1.1
Clause 7.8.1.2	Clause 7.8.1.2	Clause 7.8.1.2	Clause 7.8.1.2	Clause 7.8.1.2
Clause 7.8.1.3	Clause 7.8.1.3	Clause 7.8.1.3	Clause 7.8.1.3	Clause 7.8.1.3
Clause 7.8.1.4	Clause 7.8.1.4	Clause 7.8.1.4	Clause 7.8.1.4	Clause 7.8.1.4
Clause 7.8.2	Clause 7.8.2	Clause 7.8.2	Clause 7.8.2	Clause 7.8.2
Clause 7.8.3	Clause 7.8.3	Clause 7.8.3	Clause 7.8.3	Clause 7.8.3
Clause 7.9	Clause 7.9	Clause 7.9	Clause 7.9	Clause 7.9

(Continued)

SBCSub/D/A or SBCSub/D/C	SBCSub/A or SBCSub/C	DBSub/A or DBSub/C	ICSub/A or ICSub/C	ICSub/D/A or ICSub/D/C
Clause 7.10.1	Clause 7.10.1	Clause 7.10.1	Clause 7.10.1	Clause 7.10.1
Clause 7.10.2.1	Clause 7.10.2.1	Clause 7.10.2.1	Clause 7.10.2.1	Clause 7.10.2.1
Clause 7.10.2.2	Clause 7.10.2.2 Drafting refers to Sub-contract Works only suspended (i.e. design is not part of the sub-contract works)	Clause 7.10.2.2 Drafting refers to Sub-contract Works only suspended (but design is part of the sub-contract works)	Clause 7.10.2.2 Drafting refers to Sub-contract Works only suspended (i.e. design is not part of the sub-contract works)	Clause 7.10.2.2
Clause 7.10.2.3	Clause 7.10.2.3	Clause 7.10.2.3	Clause 7.10.2.3	Clause 7.10.2.3
Clause 7.11.1	Clause 7.11.1	Clause 7.11.1	Clause 7.11.1	Clause 7.11.1
Clause 7.11.2.1	Clause 7.11.2	Clause 7.11.2.1	Clause 7.11.2	Clause 7.11.2.1
Clause 7.11.2.2	N/A	Clause 7.11.2.2	N/A	Clause 7.11.2.2
Clause 7.11.3.1	Clause 7.11.3.1	Clause 7.11.3.1	Clause 7.11.3.1	Clause 7.11.3.1
Clause 7.11.3.2	Clause 7.11.3.2	Clause 7.11.3.2	Included in clause 7.11.3.1	Included in clause 7.11.3.1
Clause 7.11.3.3	Clause 7.11.3.3	Clause 7.11.3.3	Included in clause 7.11.3.1	Included in clause 7.11.3.1
Clause 7.11.3.4	Clause 7.11.3.4	Clause 7.11.3.4	Clause 7.11.3.2	Clause 7.11.3.2
Clause 7.11.3.5	Clause 7.11.3.5	Clause 7.11.3.5	Clause 7.11.3.3	Clause 7.11.3.3
Clause 7.11.4.1	Clause 7.11.4.1	Clause 7.11.4.1	Clause 7.11.4.1	Clause 7.11.4.1
Clause 7.11.4.2	Clause 7.11.4.2	Clause 7.11.4.2	Clause 7.11.4.2 Paragraph C.4.4 reference is to schedule 1 (not schedule 3) to the main contract conditions.	Clause 7.11.4.2 Paragraph C.4.4 reference is to schedule 1 (not schedule 3) to the main contract conditions.
Clause 7.11.5	Clause 7.11.5	Clause 7.11.5	Clause 7.11.5	Clause 7.11.5

13 Settlement of Disputes

13.1 Introduction

Despite all best efforts, from time to time disputes will arise on construction projects which cannot be resolved by the normal process of discussion and negotiation.

Because of this the JCT Sub-contracts dealt with in this book provide for four ways of resolving disputes.

These are:

- Mediation
- Adjudication
- Arbitration
- Litigation

Each of these processes are considered further within this chapter.

Against this background, in this chapter, the above matters will be dealt with, using the sub-contract clause references, etc. contained in the SBCSub/D/A or SBCSub/D/C (as appropriate). Whilst it is clearly beyond the scope of this book to review every nuance of the other sub-contract forms under consideration, the equivalent provisions (where applicable) within the SBCSub/A or SBCSub/C, the DBSub/A or DBSub/C, the ICSub/A or ICSub/C and the ICSub/D/A or ICSub/D/C are given at the table at the end of this chapter. It must be emphasised that before considering a particularly issue, the actual terms of the appropriate edition of the relevant sub-contract should be reviewed by the reader (and/or legal advice should be sought as appropriate) before proceeding with any action/inaction in respect of the sub-contract in question.

13.2 Mediation

13.2.1 Introduction

Mediation is a method of settling disputes or differences in which a third party, an independent and impartial person called a Mediator, assists both parties to reach an agreement which each party considers is acceptable.

The JCT 2011 Building Sub-contracts, First Edition. Peter Barnes and Matthew Davies.
© 2016 John Wiley & Sons, Ltd. Published 2016 by John Wiley & Sons, Ltd.

13.2.2   Advantages/Disadvantages

The main advantages of mediation, as compared with other dispute resolution processes, are its speed, flexibility and relatively low cost.

However, to be successful, both parties must be willing and able to compromise in attempting to reach a settlement.

Because mediation uses a non-adversarial approach, this often allows relationships between the parties to be maintained or rebuilt more easily.

13.2.3   Facilitative/Evaluative

In the United Kingdom, Mediators are normally facilitative, which means that they use their skills to help the parties to reach their own mutually acceptable settlement, rather than being evaluative (i.e. evaluating the liability and/or the quantum of the dispute) and then effectively imposing a decision on the parties.

13.2.4   Interests and needs

During the mediation process, Mediators encourage parties to consider their interests and needs, and not merely their strict legal rights. This is a hard concept to understand, but really goes to the heart of the mediation process, which is to reach a resolution of a dispute which both parties can accept satisfies their respective interests and needs rather than necessarily satisfying their strict legal entitlement.

13.2.5   The process

Mediation is a very flexible process which is normally adapted to suit the particular circumstances, dispute or difference, and to suit the parties' needs.

The mediation is normally conducted in a 1-day or a half-day meeting. During the day the Mediator normally holds meetings with both parties present (referred to as plenary sessions), and also holds meetings with the parties separately. During these latter meetings, the Mediator may encourage the parties to test their own case on a sensible and pragmatic basis, and consider what the alternative is (particularly in terms of legal costs, etc.) if a settlement is not reached.

13.2.6   The settlement

Any negotiated settlement between the parties in a mediation is not binding until the parties draw up a binding settlement agreement.

13.2.7   Is there an obligation to use mediation?

There is no obligation upon the parties to go to mediation at all (although if litigation is followed, the courts may expect some form of mediation to have taken place). Clause 8.1 simply says, '… if a dispute or difference arises under this Sub-contract which cannot be resolved by direct negotiations, each Party shall give serious consideration to any request by the other to refer the matter to mediation".

13.3 Adjudication

13.3.1 Introduction

Adjudication is the most common formal dispute resolution process for resolving disputes in the construction industry. The adjudication process is a means by which an independent third party (i.e. the Adjudicator) makes a (relatively) quick decision when the parties to a contract are in dispute.

13.3.2 Advantages/Disadvantages

The major advantages of adjudication are that it is relatively quick (with the entire process often being completed in 5 weeks) and inexpensive. Also, because it is backed by statute, a party to a construction contract (as defined) can refer a (crystallised) dispute to adjudication at any time. Further, the process is private and confidential, other than where enforcement proceedings become necessary.

A disadvantage is that adjudication is sometimes perceived as being 'rough justice', with the party referring the dispute seen to have an advantage as it can take many weeks/months over the preparation of its Referral Notice, whereas the party responding normally only has 7 days to produce a Response. Also, although adjudication is still a relatively inexpensive means of resolving a dispute, costs are increasing all of the time, and (unlike arbitration and litigation) a party is (virtually always) liable for his or her own legal costs in the process, irrespective of whether or not he or she is successful in the action. In other words, the normal position is that a party to an adjudication action will be required to bear his or her own legal costs in the action irrespective of whether he or she wins, loses or draws in the actual proceedings.

13.3.3 The Construction Acts

Statutory adjudication was introduced into the construction industry by way of the Housing Grants, Construction and Regeneration Act 1996 (the 'HGCRA'), and that Act of Parliament was later amended by the Local Democracy, Economic Development and Construction Act 2009 (the 'LDEDCA'), which came into effect in England and Wales on 1 October 2011.

13.3.4 The Scheme

Whilst a contract may contain its own specific and express provisions in respect of adjudication, a regulatory document containing adjudication provisions as required by the HGCRA (as amended by the LDEDCA) was produced, and that document was the Scheme for Construction Contracts (England and Wales) Regulations 1998 (Amendment) (England) Regulations 2011 (the 'Scheme').

13.3.5 JCT adjudication provisions

Rather than produce its own *ad hoc* adjudication provisions, the JCT has in effect adopted the Scheme provisions.

Article 4 of the Sub-contract Agreement states that if any dispute or difference arises under the Sub-contract, either party may refer it to adjudication in accordance with clause 8.2 of the Sub-contract conditions; clause 8.2 simply adopts the Scheme adjudication provisions with some relatively minor qualifications.

13.3.6 Qualifications to the Scheme

The first of these qualifications, as set out under clause 8.2.1 of the Sub-contract, is that the Adjudicator shall be the person named (if any) or the adjudicator-nominating body as stated in the Sub-contract Particulars under item 16 (i.e. the Royal Institute of British Architects, or the Royal Institution of Chartered Surveyors or constructionadjudicators.com; or the Association of Independent Construction Adjudicators; or the Chartered Institute of Arbitrators).

The second of these qualifications, as set out under clause 8.2.2 of the Sub-contract, says that where the dispute or difference relates to clause 3.11.3 (which relates to the opening up of work for test or inspections), and as to whether a direction issued thereunder is reasonable in all the circumstances, the Adjudicator shall (where practicable) be an individual with appropriate expertise and experience in the specialist area or discipline relevant to the direction or issue in dispute; or if the Adjudicator does not have the appropriate expertise and experience, the Adjudicator shall appoint an independent expert with the required expertise and experience to advise and report in writing on whether or not the direction under clause 3.11.3 is reasonable in all the circumstances.

13.3.7 Adjudication at any time

The parties are not obligated to go to adjudication at all, although they cannot be prevented from referring a dispute to adjudication at any time.

13.3.8 Dispute or disputes

Although the Scheme refers to the singular word 'dispute', it is accepted that this can mean a range of issues within a single dispute. Therefore, for example, a dispute about the amount to be paid in respect of a payment application would be a single dispute which may have several areas of disagreement in respect of, for example, the measured works, the variation account, and the loss and expense account.

The dispute must have crystallised, which means that a dispute only exists after a claim has been notified and rejected, although a rejection may stem from the other party's refusal to consider or answer a claim.[1] In the *Fastrack* v. *Morrison*[2] case, Judge Thornton QC held that the dispute encompassed 'whatever claims, heads of claim, issues, contentions or causes of action that are then in dispute which the referring party has chosen to crystallize into an adjudication reference'.

In addition, any dispute that has already been decided by an adjudicator following an earlier adjudication action cannot be referred to adjudication. (i.e. There is a bar on re-adjudication of the same dispute.[3])

13.3.9   Notice of Adjudication

The adjudication process is started (but may not have commenced – see section 13.3.13) by issuing a Notice of Adjudication to all of the parties to the contract (normally there are only two parties to a contract), and the Notice of Adjudication should set out briefly:

- the nature and a brief description of the dispute and the parties involved;
- details of where and when the dispute has arisen;
- the nature of the redress sought;
- the names and addresses of the parties to the contract.

The Notice of Adjudication is a very important document as it (together with the terms of the Sub-contract) sets out the jurisdiction of the Adjudicator.[4] (i.e. It sets out the boundaries of the Adjudicator's authority.) It is because of this that when asking for a decision from the Adjudicator for a sum of money, it is normal to add after the request 'or such other sum as the Adjudicator may decide'; similarly, if a time period is requested, it is normal to add the words 'or such other time period as the Adjudicator may decide', because these caveats extend the Adjudicator's jurisdiction to make a decision as to the sum that he or she may find due or the time that he or she may find due, rather than the specific sum or the specific time claimed as being due.

13.3.10   Appointment of the Adjudicator

The Adjudicator should not be appointed until after the Notice of Adjudication has been served. The Adjudicator is either the individual named under item 16 of the Sub-contract Particulars, or should be appointed by the adjudicator-nominating body as stated in the Sub-contract Particulars under item 16

---

[1] *Amec Civil Engineering Ltd* v. *The Secretary of State for Transport* [2005] CILL 2189; *Fastrack Contractors Ltd* v. *Morrison Construction Ltd* and *Impregilo UK Ltd* [2000] BLR 168.
[2] *Fastrack Contractors Ltd* v. *Morrison Construction Ltd* and *Impregilo UK Ltd[2000] BLR 168.*
[3] *Sherwood & Casson Ltd* v. *Mackenzie Engineering Ltd* [2000] CILL 1577; *VHE Construction plc* v. *RBSTB Trust Co Ltd* [2000] CILL 1592.
[4] *Fastrack Contractors Ltd* v. *Morrison Construction Ltd* and *Impregilo UK Ltd* [2000] BLR 168.

(i.e. the Royal Institute of British Architects, or the Royal Institution of Chartered Surveyors or constructionadjudicators.com; or the Association of Independent Construction Adjudicators; or the Chartered Institute of Arbitrators).

Where an Adjudicator is not named and a nominating body has not been selected, the nominating body shall be one of the bodies listed under item 16 of the Sub-contract Particulars as selected by the party requiring the reference to adjudication.

13.3.11   Adjudicator's jurisdiction

An Adjudicator has no power to rule on his or her own jurisdiction,[5] although the parties (i.e. the Referring Party and the Responding Party together) can agree to give him or her that power.[6] Even if the Adjudicator does not have the power to rule on his or her own jurisdiction, he or she can, and he or she is required to decide, in a non-binding way, upon his or her jurisdiction if called upon to do so. Any points which are to be made as to whether the Adjudicator has jurisdiction must be raised in a timely manner during the adjudication process itself,[7] because if they are not made in a timely manner, the objection may not be accepted by a court in any subsequent enforcement proceedings.

13.3.12   Adjudicators' obligations

Under paragraph 12 of the Scheme, the Adjudicator is required to act impartially in carrying out his or her duties and is required to do so in accordance with any relevant terms of the Sub-contract. The Adjudicator is to reach his or her decision in accordance with the applicable law in relation to the Sub-contract, and is to adopt a process that avoids incurring unnecessary expense.

13.3.13   The Referral Notice

The Referral Notice is effectively the Referring Party's statement of case, and should detail the particulars of the dispute and the contentions being relied upon, and should also set out a statement of the relief or remedy sought. The Referral Notice needs to be served by the Referring Party on the Adjudicator, and at the same time, on the Responding Party, no later than 7 days after the date that the Notice of Adjudication was served. The timetable for the adjudication process cannot be set out until after the Referral Notice has been served.

13.3.14   The Response

Under the Scheme, somewhat strangely, the Responding Party has no right to issue a Response. However, the Adjudicator has the power to request a Response,

---

[5] *The Project Consultancy Group* v. *The Trustees of the Gray Trust* [1999] BLR 377.

[6] *Nolan Davis Ltd* v. *Steven Catton* (2000) (unreported) QBD (TCC) No. 590.

[7] *Maymac Environmental Services Ltd* v. *Faraday Building Services Ltd* [2000] CILL 1686.

and to satisfy the requirements of natural justice (i.e. to allow both parties to have a fair opportunity to present their respective cases), an Adjudicator will almost invariably ask for a Response to be issued. Whilst there is no set period for a Response to be provided, it is commonly the case that an Adjudicator will direct that a Response should be issued by no later than 7 days after the issue of the Referral Notice.

13.3.15   The adjudication timetable

Under the Scheme there is no set timetable for the adjudication process other than that the Adjudicator shall reach his or her decision not later than either 28 days after the date of the Referral Notice (as paragraph 19(1)(a) of the Scheme), or if the Referring Party consents, 42 days after the date of the Referral Notice (as paragraph 19(1)(b) of the Scheme). Any period in excess of 42 days after the date of the Referral Notice for the Adjudicator to reach his or her decision can only apply if both the Referring Party and the Responding Party agree to the extended period (as paragraph 19(1)(c) of the Scheme).

Apart from the above long-stop dates, the Scheme does not set out any particular timetable for the submission of documents, the holding of meetings, etc., and any such timetable is set by the Adjudicator to suit the circumstances of each individual case.

13.3.16   Further submissions and the adjudication meeting

Whilst it is common for further submissions to be made after the Response, e.g. a Reply, a Rejoinder, a Rebuttal, and so on, it is much less common for an adjudication meeting to be held, and the vast majority of adjudications are decided on a documents-only basis.

In the event that an adjudication meeting is held, a party may only be represented by one person, unless the Adjudicator directs otherwise.

13.3.17   The Adjudicator's Decision

In reaching his or her decision, the Adjudicator is required to ascertain the facts and apply the law.

It must be noted that a basic principle is that 'He who asserts must prove'. In other words, a party putting forward a case must prove that case; this is what is referred to as being the 'burden of proof' on the party putting forward a claim.

In terms of the level of proof required, the Adjudicator must decide on which party's evidence is preferred on the 'balance of probabilities' principle, which is often referred to as being proved on a 'more likely than not' basis.

If the evidence of the party with the burden of proof (normally the Referring Party) is no more convincing than the evidence of the other party (the Responding Party), even if it is equally convincing, then the Referring Party (in that case) would have failed to have satisfied its burden of proof obligation on a balance of probabilities basis.

13.3.18 The Adjudicator's fees

The Adjudicator will normally inform the parties of his or her fees and expenses schedule at the time of his or her appointment.

The Adjudicator will be entitled to payment of his or her reasonable fees and expenses, and will be entitled to apportion his or her fees as he or she sees fit (although the parties will remain jointly and severally liable for the Adjudicator's fees and expenses in full) (as paragraph 9(4) of the Scheme).

More often than not, an Adjudicator will apportion his or her fees and expenses on the basis that 'costs follow the event', which means that the party that is unsuccessful in the proceedings will be responsible for the Adjudicator's fees and expenses. However, an Adjudicator is not bound by that 'convention' and is free to apportion his or her fees and expenses between the parties as he or she sees fit.

13.3.19 The parties' legal costs

The general position is that each party pays for its own legal and other costs in connection with the adjudication proceedings, irrespective of whether the party wins or loses in the adjudication action. Any application fee paid to an adjudicator-nominating body is normally considered to be part of the costs of the Referring Party in going to adjudication.

If both parties expressly give the Adjudicator the jurisdiction to apportion the parties' own legal and other costs, then the Adjudicator may do so. However, this situation very rarely arises in practice.

13.3.20 The effect of the Adjudicator's Decision

Once an Adjudicator has reached his or her Decision, he or she becomes 'functus officio' (i.e. his or her mandate has expired because he or she has completed the task for which he or she was appointed). The only exception to this is that the Adjudicator may either on the application of either party or upon his or her own initiative correct his or her Decision so as to remove a clerical or typographical error arising by accident or omission, with any such correction being made within 5 days of the delivery of the Decision to the parties (as paragraph 22A of the Scheme).

The above provision only applies to the correction of clerical or typographical errors; it does not permit an Adjudicator to correct his or her Decision to rectify an error in fact or in law. Generally, provided that an Adjudicator answers the correct question, the parties are bound by any error of fact or law contained within the Decision.

The Decision of the Adjudicator is binding on the parties, and the parties shall comply with it until the dispute is finally determined by legal proceedings, by arbitration (if the contract provides for arbitration or the parties otherwise agree to arbitration) or by agreement between the parties.

13.3.21   Enforcement of the Adjudicator's Decision

Even though an Adjudicator's Decision is what is referred to as 'temporarily binding[8]' upon the parties, in the vast majority of cases, and anecdotally, the parties comply with the Adjudicator's Decision as though it was permanently binding upon them.

However, where a party does not comply with an Adjudicator's Decision, the other party may pursue enforcement proceedings through the courts in a fast-track process.

If it is successful in those enforcement proceedings, the successful party can normally recover its reasonable legal costs in making the application. However, if it is unsuccessful, then it may be liable for the other party's legal costs. It should be noted that the recovery of 'reasonable' legal costs is usually significantly less than the actual legal costs incurred.

Although the courts are generally very supportive of the adjudication process, not all Adjudication Decisions are enforced. The principal reasons why an Adjudicator's Decision may not be enforced are the lack of jurisdiction, procedural unfairness, breach of natural justice, actual or perceived bias, or an incomplete Decision.

13.3.22   Final determination of the dispute

It should be noted that when the dispute is finally determined by legal proceedings or by arbitration (as applicable), the Adjudicator's Decision is not open for review or appeal; the dispute is simply considered again with a replacement process being used (i.e. either arbitration or litigation as appropriate) as though the Adjudicator's Decision had not taken place.

No appeal can be made against an Adjudicator's Decision. If either party is not prepared to accept the Adjudicator's Decision as being a final decision on the dispute, it can refer the dispute to either legal proceedings or to arbitration, whichever is applicable, but in the interim, the Adjudicator's Decision must be complied with.

13.4   Arbitration

13.4.1   Introduction

Arbitration may be defined as a private procedure for settling disputes whereby a dispute between parties is decided judicially by an impartial tribunal (normally an individual, even though it could be in theory, but in the domestic market hardly ever is in practice, a panel of individuals) either selected by the parties or appointed for that purpose.

---

[8] *Macob Civil Engineering Ltd* v. *Morrison Construction Ltd* (1999) 1 BLR 93; *Herschel Engineering Ltd* v. *Bream Property Ltd* [2000] BLR 272.

An individual so appointed is referred to as the Arbitrator, whilst his or her decision is referred to as an Award. An Arbitrator's Award is legally binding on all the parties to the arbitration proceedings to whom it is addressed.

Arbitration in England is governed by the Arbitration Act 1996; the arbitration process applicable in a particular situation is also governed by any applicable arbitration 'rules' stipulated in the Sub-contract in question.

13.4.2   Advantages/Disadvantages

The major advantages of arbitration are that it can (particularly if a robust Arbitrator is appointed) be quicker and less expensive than litigation (although the Arbitrator and the hearing room must be paid for unlike a Judge and the court, which are not charged); the dispute is decided with a 'private' judge (i.e. the Arbitrator), who usually has some knowledge and experience of the issues involved; the process is private and confidential (to a large degree); and the parties (with the Arbitrator) can have an input into the process and the timetable to be followed. Also, there are very few opportunities for an appeal to the Arbitrator's Award to be made.

The disadvantages are that if the Arbitrator is not robust, the process can become considerably prolonged, because the parties often cannot agree upon anything, and the costs can increase to an alarming degree. It is easy to take the first step in arbitration, but it is often nowhere near so easy to get out of the quicksand formed by the potential cost liability afterwards, and frequently, the cost issue itself is an obstacle to a settlement being reached by the parties. A further issue may be that the Arbitrator may not be an expert on the law, and therefore, if important legal issues are at stake, the Arbitrator may not be able to properly deal with the issues.

13.4.3   How is arbitration incorporated into the JCT Sub-contracts?

Arbitration is incorporated by way of article 5.

When the Sub-contract is executed, item 2 of the Sub-contract Particulars (which relates to arbitration) should be marked so that article 5 either does or does not apply:

- If article 5 is marked to apply, then *arbitration will apply* and legal proceedings will not apply.
- If article 5 is marked not to apply, then *arbitration will not apply* and legal proceedings will apply;
- If article 5 is not marked at all, then the default position is that *arbitration will not apply* and legal proceedings will apply.

Therefore, as the default position is that disputes and differences under this Sub-contract are to be finally determined by legal proceedings. If the parties wish arbitration to apply, they must mark Sub-contract Particulars item 2 so that article 5 applies.

Where arbitration does apply, clauses 8.3–8.8 of the Sub-contract apply.

Clause 8.8 of the Sub-contract makes it plain that the provisions of the Arbitration Act 1996 shall apply to any arbitration under the Sub-contract 'wherever the same, or any part of it, shall be conducted'. This means that even if the Sub-contract works were carried out outside of the jurisdiction of England, the provisions of the Arbitration Act 1996 would still apply.

13.4.4   If arbitration is incorporated into the JCT Sub-contracts, can the parties ignore this and refer a dispute to court?

If all of the parties to the Sub-contract agree not to refer a dispute to arbitration (even though that was the process selected in the Sub-contract) but agree to refer the dispute to litigation, then the dispute can be referred to court.

However, if one party attempts to refer a dispute to court in the case where an arbitration agreement exists, section 9(1) of the Arbitration Act 1996 makes it clear that the other party may apply to the court in which the proceedings have been brought to stay the proceedings in favour of arbitration.

Section 9(4) of the Arbitration Act 1996 states that upon such an application being made, 'the court shall grant a stay unless it is satisfied that the arbitration agreement is null and void, inoperative, or incapable of being performed'.

However, if a step is taken in the court process before a stay for arbitration is sought, the court will most likely say that the required stay will not be granted.

13.4.5   If a Sub-contract ceases to exist, does the arbitration agreement also cease to exist?

It should be noted that although the parties may incorporate the arbitration agreement within the JCT Sub-contract, the arbitration agreement is in legal terms separable from and survives the termination of the Sub-contract. It is as though the arbitration agreement is a separate contract between the parties.

In this regard, section 7 of the Arbitration Act 1996 states that:

'Unless otherwise agreed by the parties, an arbitration agreement which forms or was intended to form part of another agreement (whether or not in writing) shall not be regarded as invalid, non-existent or ineffective because that other agreement is invalid, or did not come into existence or has become ineffective, and it shall for that purpose be treated as a distinct agreement'.

13.4.6   What are the rules for the conduct of the arbitration?

Clauses 8.3 of the Sub-contract makes it plain that the JCT 2011 edition of the Construction Industry Model Arbitration Rules (CIMAR) will apply. If any amendments to those rules have been issued at the time when the proposed arbitration is due to commence, then, provided that the Arbitrator is provided with a joint notice from both parties saying that they wish those amended rules to apply, then those amended rules will apply.

Also, as noted above, clause 8.8 of the Sub-contract makes it plain that the provisions of the Arbitration Act 1996 shall also apply.

13.4.7 What are the Construction Industry Model Arbitration Rules (CIMAR)?

The Arbitration Act 1996 confers wide powers on the arbitrator unless the parties have agreed otherwise, but leaves detailed procedural matters to be agreed between the parties or, if not so agreed, to be decided by the arbitrator.

To avoid problems arising, it is generally advisable to agree as much as possible of the procedural matters in advance, and where arbitration applies in respect of the JCT Sub-contracts, this is achieved by the incorporation of the CIMAR, which are clearly written arbitration rules.

13.4.8 What is the procedure to be followed under CIMAR?

CIMAR Rule 5.1 makes it clear that the Arbitrator shall decide all procedural and evidential matters subject to the right of the parties to agree any matter.

13.4.9 How is the arbitration commenced?

CIMAR Rule 2.1 says that arbitral proceedings are begun in respect of a dispute when one party serves on the other a written notice of arbitration identifying the dispute and requiring him or her to agree to the appointment of an Arbitrator.

13.4.10 What is a notice of arbitration?

A notice of arbitration identifies (briefly) the dispute between the parties and asks the other party to agree to the appointment of an Arbitrator. It is quite common for a list of three names of prospective Arbitrators to be included in the notice of arbitration.

13.4.11 The appointment of the Arbitrator

Clause 8.4.1 of the Sub-contract states that:

'the Arbitrator shall be an individual agreed by the Parties or, failing such agreement within 14 days (or any agreed extension of that period) after the notice of arbitration is served, appointed on the application of either Party in accordance with Rule 2.3 by the person named in the Sub-contract Particulars'.

The Arbitrator selected by the parties should, of course, be somebody with the necessary expertise and experience to deal with the dispute or difference being referred to arbitration.

There is no facility for an Arbitrator to be named in the Sub-contract, so if the parties cannot agree upon an Arbitrator within 14 days (or such longer period as agreed by the parties) of the notice of arbitration being served, then an application should be made to an arbitrator-appointing body set out under item 16 of the Sub-contract Particulars.

The parties should agree upon one of a choice of the three appointing bodies listed under item 16 of the Sub-contract Particulars (i.e. the President or a Vice-President of the Royal Institute of British Architects, or of the Royal Institution of Chartered Surveyors, or of the Chartered Institute of Arbitrators) by deleting two of the listed bodies. If the parties fail to select an appointing body when executing the Sub-contract, then the default appointing body is deemed to be the President or a Vice-President of the Royal Institution of Chartered Surveyors.

13.4.12   What procedure should the Arbitrator follow?

CIMAR Rule 6.1 states that as soon as he or she is appointed, the Arbitrator must consider the form of procedure which is most appropriate to the dispute; CIMAR Rule 6.2 says that, for this purpose, the parties shall, as soon as practicable after the Arbitrator is appointed, provide to each other and to the Arbitrator:

- a note stating the nature of the dispute with an estimate of the amounts in issue;
- a view as to the need for and length of any hearing;
- proposals as to the form of procedure appropriate to the dispute.

Thereafter, the Arbitrator should convene a procedural meeting (pursuant to CIMAR Rule 6.3) (normally referred to as a Preliminary Meeting) so that he or she may issue directions as to the appropriate procedure to be followed in the circumstances of the dispute in question. Such a meeting is not necessary if both parties agree that a meeting does not need to be held (CIMAR Rule 6.6).

The directions of the Arbitrator may be that:

- CIMAR Rule 7 (a short hearing);
- CIMAR Rule 8 (a documents-only arbitration);
- CIMAR Rule 9 (the full arbitration procedure); or
- any combination of parts of the above rules, or any other procedure that the Arbitrator considers to be appropriate, will be followed.

Under all three CIMAR Rules referred to above, the parties exchange statements of claim and statements of defence, together with copies of documents and witness statements on which they intend to rely. Rule 8 is a documents-only process, so there is no hearing held; Rule 7 allows for a short hearing; and Rule 9 is for the full arbitration process including all necessary hearings.

13.4.13   Joinder provisions

Clause 8.4.2 of the Sub-contract says that where two or more related arbitral proceedings in respect of the main contract works or the Sub-contract works fall under separate arbitration agreements, CIMAR rules 2.6, 2.7 and 2.8 shall apply.

CIMAR rules 2.6, 2.7 and 2.8 impose duties on persons having the function of appointing Arbitrators to give consideration as to whether the same or a different Arbitrator should be appointed where two or more related arbitral proceedings are commenced.

In such a situation:

- CIMAR Rule 2.6 says that the same Arbitrator should be appointed unless sufficient grounds are shown for not doing so;
- CIMAR Rule 2.7 says that where an Arbitrator has already been appointed in respect of one arbitral proceeding, due consideration should be given to the appointment of that same Arbitrator in respect of a related arbitral proceedings;
- CIMAR Rule 2.8 says that two different appointers must also give due consideration for the appointment of the same Arbitrator for related arbitral proceedings.

Clause 8.4.3 of the Sub-contract says that after an Arbitrator has been appointed, either party may give a further notice of arbitration to the other party and to the Arbitrator referring any other dispute which falls under article 5 to be decided in the arbitral proceedings, and CIMAR Rule 3.3 shall apply.

CIMAR Rule 3.3 makes it clear that if a party does not consent to the dispute being referred through this joinder provision, the Arbitrator may order either that the other dispute should be referred to and consolidated with the same arbitral proceedings, or that the other dispute should not be so referred.

It should be noted that the courts will generally give a purposeful interpretation to joinder provisions that allow for the same Arbitrator to be appointed in related disputes to avoid a multiplicity of proceedings, which might lead to excessive costs and inconsistent judgments.[30]

13.4.14 What are the powers of the arbitrator?

As noted above, the powers of the Arbitrator are derived from the Arbitration Act 1996 and the CIMAR.

Clause 8.5 of the Sub-contract says that the Arbitrator's powers include that the Arbitrator may:

- rectify the Sub-contract so that it accurately reflects the true agreement made by the parties;
- direct such measurements and/or valuations as may in his or her opinion be desirable in order to determine the rights of the parties;
- ascertain and award any sum which ought to have been the subject of or included in any certificate, and to open up, review and revise any certificate, opinion, decision, requirement or notice;
- determine all matters in dispute which shall be submitted to him or her in the same manner as if no such certificate, opinion, decision, requirement or notice had been given.

13.4.15 What is the effect of the arbitrator's award?

Clause 8.6 of the Sub-contract states that, subject to clause 8.7, the Arbitrator's award shall be final and binding on the parties.

The only exceptions (as noted in clause 8.7) being that the parties agree (pursuant to section 45(2)(a) and section 69(2)(a) of the Arbitration Act 1996) that either party may (upon notice to the other party and to the Arbitrator):

1. apply to the courts to determine any question of law arising in the course of the reference; or
2. appeal to the courts on any question of law arising out of an award made in an arbitration under the arbitration agreement.

In addition to the above, challenges to the Award may also be made on the grounds of lack of substantive jurisdiction (section 67 of the Arbitration Act 1996), and on the grounds of serious irregularity (section 68 of the Arbitration Act 1996).

In respect of all of the challenges or appeals, section 70 of the Arbitration Act 1996 makes it clear that an application to the court cannot be made until any available arbitral processes of appeal or review, or any available recourse under section 57 (correction or award or additional award) of the Arbitration Act 1996 have been exhausted.

If a party takes part, or continues to take part, in arbitral proceedings without making, either forthwith or within such time as is allowed by the arbitration agreement or the tribunal, or by the relevant provision of the Arbitration Act 1996, an objection that the tribunal lacks substantive jurisdiction, that the proceedings have been improperly conducted, that there has been a failure to comply with the arbitration agreement or that there has been any other irregularity affecting the tribunal or the proceedings, he or she may not raise that objection later, before the tribunal or the court, unless he or she shows that, at the time he or she took part in the proceedings, he or she did not know and could not with reasonable diligence have discovered the grounds for the objection (as section 73 of the Arbitration Act 1996).

13.4.16 Who pays for the costs of arbitration?

In respect of costs, the general principle is that costs should be borne by the losing party.

Therefore, subject to any agreement between the parties, the Arbitrator has a very wide discretion in awarding which party should bear what proportion of the costs of the arbitration. The costs of the arbitration include both the parties' costs and the tribunals' fees and expenses.

In arbitration, costs may be awarded either on:

- an indemnity basis (i.e. reasonable costs reasonably incurred, with any doubt as to any costs that were reasonable or were reasonably incurred being resolved in favour of the receiving party); or on
- a standard basis (i.e. reasonable costs reasonably incurred proportionate to the matters in issue, with any doubt as to any costs that were reasonable or were reasonably incurred being resolved in favour of the paying party).

The standard basis is used in the majority of cases, and under the standard basis, a successful party may (on average) only recover about 70–75% of his or her total costs incurred in the arbitration.

The 25–30% balance of the successful party's total costs is known as the non-recoverable costs, and this can be a significant factor when, on a large arbitration, a party's costs in the arbitration may run to tens of or even hundreds of thousands of pounds.

The question of costs in arbitration is normally greatly influenced by 'Calderbank' offers made by the defending party.

A Calderbank offer is the arbitral equivalent of making a Civil Procedure Rules (CPR) part 36 offer in the courts. When considering costs, the question for the Arbitrator is whether the claimant has achieved more by rejecting the offer and going on with the arbitration than he or she would have achieved if he or she had accepted the offer. If the claimant fails to achieve more than he or she would have by accepting the offer, then he or she is likely to have an award of costs made against him or her.

The Arbitrator will be entitled to charge fees and expenses and will apportion those fees between the parties as he or she sees fit (normally on the basis that costs will follow the event). Irrespective of how the Arbitrator apportions his or her fees and expenses, the parties are jointly and severally liable to the Arbitrator for the fees and expenses incurred.

13.5   Litigation

13.5.1   What is litigation?

Legal proceedings are often referred to as litigation and this involves taking a dispute for resolution through the civil courts.

13.5.2   Advantages/Disadvantages

The major advantages of litigation are that the judge is (or at least should be) an expert on the law, and the CPR require judges to manage their caseloads effectively and efficiently, and also encourage pre-action settlement through the use of a Pre-Action Protocol. Joinder provisions are simpler, and the cost of the court and of the judge is not charged. If a judgment is obtained which is not acceptable to a party, there is an appeals process (to a higher court) that can usually be followed.

The disadvantages are that often even specialist judges know relatively little about the details of construction work and often rely upon experts (who themselves may not always be capable of dealing with the case in question) to assist them. Also, of course, the judge cannot be 'chosen' and there are varying ability levels even amongst specialist judges. Further, cases can sometimes take many years to be finally disposed of (particularly when appeals are taken into account) and the lengthy timescale can make the legal process prohibitive in terms of costs.

13.5.3   What is the litigation procedure?

Procedure in the civil courts is governed by statutory CPR. The CPR are contained in separate parts, which set out the specific procedural rules to be followed in specific situations (e.g. service of documents, statements of case, summary judgment, disclosure, evidence, costs).

The CPR implement recommendations of the Woolf report (*Access to Justice*, 1994), and seek to improve the speed, efficiency and accessibility of the civil court procedure.

CPR part 1 sets out the overriding objective of CPR and that is to enable the court to deal with cases justly (i.e. to ensure, so far as is practicable, that the parties are on an equal footing, to save expense, to deal with the case in a way that is proportionate and to ensure that the case is dealt with expeditiously and fairly).

13.5.4   What is the litigation process?

The basic steps involved in a civil action in the Queen's Bench Division (where most disputes in the construction industry would be dealt with) are:

■ The action is begun by the claimant issuing and serving a claim form.
■ The defendant must then serve a defence or an acknowledgement of service.
■ There may be a counterclaim from the defendant and reply from the claimant, in which the cases are defined.
■ There is then a procedure of disclosure of documents and inspection of documents.
■ Witness statements are usually prepared.
■ Eventually, a trial is held, which culminates in a judgment.

The process leading up to the trial is known as the interlocutory proceedings.

At the commencement of any legal proceedings, the court will allocate the action to one of the three tracks on the following basis:

■ Small-claims track – appropriate for claims not exceeding £10,000. The small-claims track is a largely documents-only process.
■ Fast track – appropriate for most cases where the amount claimed is over £10,000 but does not exceed £25,000. The fast track is used where the trial is not expected to exceed 1 day.
■ Multi-track – appropriate for all other cases.

In certain situations, a claimant may apply to a court for judgment on his or her claim on the ground that there is no (or no sufficient) defence. This is known as summary judgment, and if the defendant is unable to satisfy the court that there is an issue which ought to be tried, the claimant will be entitled to immediate judgment (i.e. summary judgment) on the claim or the part of the claim in question. This application will only be successful where a defendant has no (or no sufficient) defence.

13.5.5   Who pays for the costs of litigation?

A successful party in legal proceedings is entitled to an order for payment of his or her legal costs by the loser (and the loser must also pay his or her own legal costs).

However, the award of costs is at the discretion of the courts, and because of this and because costs are normally awarded on a standard basis (i.e. any doubt as to whether costs were reasonably incurred or were reasonable and proportionate in amount to be resolved in favour of the paying party), a party that is entirely successful in its claim may only recover somewhere in the region of 70–75% of its actual legal costs.

13.5.6   Is there any protection against the costs of litigation?

A degree of protection against the costs of litigation may be obtained by a party making an offer of settlement.

CPR part 36 allows a defendant to make an offer to settle, and (in simple terms) if the claimant does not exceed the amount of that offer at the end of the trial (even though the claimant may have won the action), then he or she is liable to pay both his or her own legal costs and the defendant's legal costs from the date of the notification of the CPR part 36 offer.

13.5.7   Must a party use ADR (Alternative Dispute Resolution) before going to court?

There is no strict obligation on a party to use ADR (Alternative Dispute Resolution) before going to court; however, if a party does not use ADR, then the consequence could be that a cost sanction could be imposed, because in deciding what order (if any) to make about costs, the court must have regard to all of the circumstances, including the conduct of the parties. The impact of ADR, in respect of construction Sub-contracts, is lessened by the pre-action protocol for construction and engineering disputes.

13.5.8   What is the pre-action protocol for construction and engineering disputes?

In respect of disputes in construction, the CPR incorporates a pre-action protocol for construction and engineering disputes.

The objectives of the protocol are:

1. to encourage the exchange of early and full information about the prospective legal claim;
2. to enable parties to avoid litigation by agreeing a settlement of the claim before commencement of proceedings;
3. to support the efficient management of proceedings where litigation cannot be avoided.

13.5.9   What is the procedure of the pre-action protocol for construction and engineering disputes?

The procedure of the pre-action protocol for construction and engineering disputes is as follows:

- The claimant issues to the defendant a letter of claim.
- The defendant must acknowledge the letter of claim within 14 days (if the defendant does not acknowledge the letter of claim within 14 days, the claimant is entitled to commence proceedings without further compliance with the protocol).
- The defendant must then issue a response and, if appropriate, a counterclaim within 28 days from the date of receipt of the letter of claim (the period of 28 days can be extended by agreement between the parties up to 3 months).
- The claimant is to issue a response to any counterclaim within the equivalent period allowed to the defendant to respond to the letter of claim.
- After the exchange of the above submissions, the parties should meet at a pre-action meeting in an attempt to narrow the issues in dispute, and to agree whether and what form of ADR procedure would be more suitable to settle the outstanding disputes, rather than litigation.

13.5.10   How is litigation (legal proceedings) incorporated into the JCT Sub-contracts?

Legal proceedings are incorporated into the Sub-contract by way of article 6.

When the contract is executed, item 2 of the Sub-contract Particulars in respect of article 5 (relating to arbitration) should be marked either to apply or not to apply:

- If article 5 is marked to apply, then arbitration will apply and legal proceedings will not apply.
- If article 5 is not marked either to apply or not to apply, article 5 will be deemed *not* to apply, and disputes and differences will be determined by legal proceedings.

Therefore, the default position is that disputes and differences under this Sub-contract are to be finally determined by legal proceedings.

If the parties wish the legal proceedings to be decided by a jurisdiction other than the English courts, then the appropriate amendment must be made to article 6.

13.6   Equivalent sub-contract provisions

In the table below, the equivalent sub-contract provisions to that within the SBCSub/D/A or SBCSub/D/C in the text above are listed for the SBCSub/A or SBCSub/C, DBSub/A or DBSub/C, ICSub/A or ICSub/C and ICSub/D/A or ICSub/D/C.

SBCSub/D/A or SBCSub/D/C	SBCSub/A or SBCSub/C	DBSub/A or DBSub/C	ICSub/A or ICSub/C	ICSub/D/A or ICSub/D/C
Article 4	Article 4	Article 4	Article 4	Article 4
Article 5	Article 5	Article 5	Article 5	Article 5
Article 6	Article 6	Article 6	Article 6	Article 6
Clause 8.1	Clause 8.1	Clause 8.1	Clause 8.1	Clause 8.1
Sub-contract Particulars Item 2	Sub-contract Particulars Item 2	Sub-contract Particulars Item 2	Sub-contract Particulars Item 2	Sub-contract Particulars Item 2
Sub-contract Particulars Item 16	Sub-contract Particulars Item 15	Sub-contract Particulars Item 16	Sub-contract Particulars Item 14	Sub-contract Particulars Item 14
Clause 3.11.3	Clause 3.11.3	Clause 3.11.3	Clause 3.10	Clause 3.10
Clause 8.1	Clause 8.1	Clause 8.1	Clause 8.1	Clause 8.1
Clause 8.2	Clause 8.2	Clause 8.2	Clause 8.2	Clause 8.2
Clause 8.2.1	Clause 8.2.1	Clause 8.2.1	Clause 8.2.1	Clause 8.2.1
Clause 8.2.2	Clause 8.2.2	Clause 8.2.2	Clause 8.2.2	Clause 8.2.2
Clauses 8.3–8.8	Clauses 8.3–8.8	Clauses 8.3–8.8	Clauses 8.3–8.8	Clauses 8.3–8.8
Clauses 8.4.1–8.4.3	Clauses 8.4.1–8.4.3	Clauses 8.4.1–8.4.3	Clauses 8.4.1–8.4.3	Clauses 8.4.1–8.4.3
Clause 8.5	Clause 8.5	Clause 8.5	Clause 8.5	Clause 8.5
Clause 8.6	Clause 8.6	Clause 8.6	Clause 8.6	Clause 8.6
Clause 8.7	Clause 8.7	Clause 8.7	Clause 8.7	Clause 8.7

Table of Cases

The JCT 2011 Building Sub-contracts, First Edition. Peter Barnes and Matthew Davies.
© 2016 John Wiley & Sons, Ltd. Published 2016 by John Wiley & Sons, Ltd.

Table of Statutes and Regulations

The JCT 2011 Building Sub-contracts, First Edition. Peter Barnes and Matthew Davies.
© 2016 John Wiley & Sons, Ltd. Published 2016 by John Wiley & Sons, Ltd.

Index

The JCT 2011 Building Sub-contracts, First Edition. Peter Barnes and Matthew Davies.
© 2016 John Wiley & Sons, Ltd. Published 2016 by John Wiley & Sons, Ltd.

Printed and bound by CPI Group (UK) Ltd, Croydon, CR0 4YY

27/10/2024

14580357-0001